C R Y S T A L S

CRYSTALS

JENNIE HARDING

CHARTWELL
BOOKS

This edition published in 2016 by
CHARTWELL BOOKS
an imprint of The Quarto Group
142 West 36th Street, 4th Floor
New York, New York 10018
USA

First published in the UK in 2007

© 2017 Quarto Publishing plc

ISBN-13: 978-0-7858-3398-7

Printed in Singapore COS092020

11

This book was conceived, designed, and produced by
Ivy Press
an imprint of The Quarto Group
The Old Brewery, 6 Blundell Street
London N7 9BH, United Kingdom
T (0)20 7700 6700 **F** (0)20 7700 8066
www.QuartoKnows.com

Creative Director Peter Bridgewater
Publisher Jason Hook
Editorial Director Caroline Earle
Project Editors Hazel Songhurst, Mary Todd
Art Director Sarah Howerd
Senior Designer Suzie Johanson
Design JC Lanaway
Photographer Stephen Marwood
Picture Research Katie Greenwood, Elizabeth Carlé

Contents

l as the rainbow and sparkling with light, crystals, pre
and minerals have fascinated people since the beginr
nusual and precious stones were picked up or dug out
y our ancestors to be worn as ornaments or carried as
ection. Throughout history, beautiful stones have bee
ged, fought over, collected, used as healing tools, or w
te jewelry, and today interest in crystals and minerals
ular as ever.

are crystals so fascinating? Perhaps because of their
y—they come in so many colors, shapes, and sizes; pe
e they all come from the earth itself, miraculously form
t, and they remind us of the mystery and beauty of ou
s because they are rare and beautiful and seem to reac
their presence. When people come into contact with c
ant to touch them, hold them, and feel their shape, ser
. Examples such as soft pink rose quartz, sparkling yel
en moss agate, or deep blue lapis lazuli all have uniqu

The World of Color

Color has a profound effect on how we react to our surroundings. If we wake up to gray skies and rain for several days, it can be depressing. Then if suddenly there is a bright blue sky outside, the leaves on the trees look very green, the sunlight brings out the colors of flowers, and our moods begin to change as a result. All this is due to the way we perceive color through the eyes, our organs of sight. The eye is a sophisticated light box that allows us to receive and interpret different shades of color. Light frequencies

also have very real effects on the brain, influencing our internal chemistry. The gift of sight and the ability to enjoy colors is crucial to the way we perceive and react to the world around us. In this chapter, we will explore in detail how the eye reacts to color, breaking down the seven colors of the rainbow into more varied shades, and find out how color healing uses different shades for specific effects on body and mind. This will help us connect later in the book with all the variations in colors present in crystals and minerals.

What is color and how do we see it?

The eyes allow color to be perceived and interpreted by the brain. They account for approximately 70 percent of our total sensory perception, not only of color but also perspective and distance. Their ability to filter and interpret light contributes to the majority of our everyday experiences, because what we see influences how we feel and react in almost every moment.

STRUCTURE OF THE EYES

Each eye is a small, roughly spherical structure that is about 1 inch (3 centimeters) in diameter. Only around one-sixth of its surface is exposed to the outside—this is the "eye" we see in a person's face. The remainder of the eye's surface is set back inside the skull and is protected by a bony socket. The eyes connect to optical nerves that relay the impressions of light back to the brain, particularly an area at the back of the brain called the visual cortex, where visual impressions are interpreted.

Inside each eyeball are a number of important structures. There is a lens that becomes thinner when you look into the distance or thicker when you need to see close up. The iris, the colored part of the eye, is actually a muscle that changes the size of the pupil in order to control the amount of light entering the eye. At the back of the eye is the retina, a layer with two types of light-sensitive cells called rods and cones. The rods help us see in darkness and distinguish black, white and shades of gray; the cones enable us to see red, green, and blue, and are better suited to detecting fine detail in daylight.

EYE STRUCTURE

The eye is a small but miraculous anatomical structure that lets in light and converts it to images thanks to cells in the retina.

LIGHT AND THE EYES

The main rays of the color spectrum appear in a rainbow—red, orange, yellow, green, blue, indigo, and violet—when sunlight splits white light into colors by shining through rain droplets. There are other bands of light our eyes cannot see, such as ultraviolet rays from sunlight and infrared rays used in security systems or in the remote control of your television. Electric light is artificially created, and operates mainly in the visible part of the spectrum.

ABOVE *Being able to see color is one of the most miraculous gifts of human physiology; it adds immense variety to our lives.*

Over millions of years of evolution, our eyes and brains have become used to interpreting color. Reds, oranges, and yellows appear warm to us, and they have a stimulating or energizing effect; greens, blues, and violets are cooler hues, with a more soothing effect. Although we live in artificial environments, we still long for outdoor landscapes filled with blue sky, and we enjoy the colors of flowers balanced by green leaves. For thousands of years before the invention of electric light, the rhythms of our lives were far more governed by the presence or absence of natural light, and the effects of light on our eyes dictated how we lived.

An expanded color wheel

Traditional representations of the color wheel show seven shades found in the rainbow—red, orange, yellow, green, blue, indigo, and violet. However, close examination of an actual rainbow in the sky shows that the main bands of color are not distinct; they merge into each other. So, for example, from deep green the color slowly changes to turquoise—blue and green mixed together—to aquamarine, to pale blue, and then into the pure blue shades. This gradual bleed of color creates a whole new palette of more subtle shades, all of which are represented in the range of available crystals.

THE EXPANDED SPECTRUM

Twenty shades are represented here, including tones such as brown, which is linked very much to the physical earth, and silver-gray, exemplified by the metal silver, yellow-gold to include the metallic element gold, and even black, the opposite to white in the spectrum. Black is a very important resonant shade in healing, and does not always have negative associations, being seen rather as a cleansing vibration. White is the shade that encompasses all the colors of the spectrum.

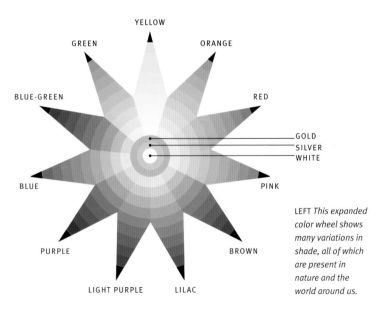

LEFT *This expanded color wheel shows many variations in shade, all of which are present in nature and the world around us.*

If you paint a circle with the seven true rainbow colors and then spin it fast enough, the shades will all merge to white.

This expanded spectrum of color is exciting because it provides an opportunity to distinguish more clearly between subtle differences in shades. Light purple as a shade, for instance, is exemplified by pale amethyst crystals, and lilac is similar but with a hint of pink—lavender quartz is a good example, a softly lilac-colored variation of rose quartz. In the blue-and-green area of the spectrum, there are many variations of color, illustrated in nature by an image of the sky and the ocean together. Earth when seen from space shows that our blue planet actually contains all the shades from turquoise to aquamarine to deep blue.

Some people have a more restricted impression of colors, not because they cannot see them but because their minds have not had the opportunity to learn to appreciate the subtle shades. Working with crystals and studying them closely opens the eyes to new levels of vibration in the form of color. Once these are recognized in the context of stones they will be noticed elsewhere. Vision, the interpretation of what we sense through the eyes, can be trained to expand into new areas. Allow the crystals displayed in the Directory to bathe your eyes with new color frequencies.

13

BELOW *The visible rainbow contains the seven major colors of the spectrum as well as many other delicate shades that blend into each other.*

Color healing

Every day, thanks to our sun, living on the earth means that we are literally bathed in full-spectrum light, that is, the colors we can see plus invisible rays, such as ultraviolet and infrared. Scientific research has shown that we need to be exposed to full daylight for at least twenty minutes each day to maintain an ideal internal balance of the body's chemistry. Inside the brain are special centers that respond to light—including the pineal and pituitary glands that control the body's hormonal balance.

Lack of exposure to full daylight can have detrimental effects on health. Modern life is lived mostly indoors, away from full natural sunlight, trapped under artificial electric lighting. Medical conditions—including depression, poor sleep, low energy, and mood swings—can be caused by lack of sunlight, and this is sometimes called Seasonal Affective Disorder (SAD). Exposure to full-spectrum light for several hours daily is enough to correct many of these symptoms.

GARNET

THE EFFECTS OF COLORED LIGHT

Color healing directs light in different spectrum colors directly onto or into the body. Application of specific colors, either directly with equipment or visualized by the therapist, can bring about speedy and effective rebalancing of the body's energies. Red light has both a warming and stimulating effect and can raise blood pressure. Orange shades are energizing; yellow, which simulates the rise of the sun in the sky, awakens mental awareness. Green light has a soothing effect, calming the breathing. Relaxation responses increase in the presence of blue shades of light, and as these shades darken they simulate night falling and bring on a sense

SUNSTONE

GOLD

AMAZONITE

CELESTITE

of sleepiness. All colors of the spectrum have their own vibrational frequency, and when they are directed at the body they are absorbed, leading to a wide variety of effects.

ABOVE Natural scenery full of water, trees, sand and rocks has a soothing effect on our minds, bodies and spirits.

Some therapists use color lamps or machines to project light through crystals to shine rays of true color onto the body. These have been scientifically shown to ease many physical problems. Emerald green light, for example, can help to heal broken bones and regenerate tissue, and ruby red light can raise body temperature and increase the metabolic rate. Blue light can have a healing effect on burns or damaged skin, while psychological or mental issues benefit from antidepressant orange and apricot shades, and pink lowers aggression.

German scientist Fritz-Albert Popp has suggested that human beings are essentially beings of light. His research has proved that humans actually emit light at certain frequencies; he believes that our whole metabolism depends on light, and that the use of light is set to become an important healing tool in the twenty-first century.

Crystals and color

Many people ask how crystals work or how they have effects on the body and mind. For almost a century, science has shown that everything is made up of energy—a rock, a tree, a river, a human being. The same building blocks created the earth and all the creatures on it, and the only difference is how those blocks are assembled. Therefore it follows that since we are all made up of similar elements, it is possible to have a connection with everything around us, and crystals are another part of that picture. Some people—the Native Americans, for example—still believe that everything is alive, even rocks. If you are drawn to a certain crystal or stone it is because it resonates with you.

We have seen that light, split into the colors of the spectrum, can have noticeable effects on body and mind. Light is energy. Light projected through crystalline structures can be even more powerful, as used in some lasers, in which rubies or sapphires are used to create highly concentrated beams of light particles. Crystals reflect different colors because of their mineral composition; they attract us because of that color, their shape, and their beauty. In this way, it can be said that the mineral kingdom communicates with us.

HOW CRYSTAL COLOR IS USED

Over thousands of years of handling, wearing, and use, different crystals and minerals have become associated with particular healing effects. One example is bloodstone, a dark green agate flecked with red. The dramatic appearance of those red specks reminded people of blood, so the stone was carried by warriors because it was said to stop blood flowing from wounds. This could be regarded as mere superstition, but sometimes a powerful belief can bring about real results.

BLOODSTONE

ROSE QUARTZ

These days people's experiences of crystals are very individual. For example, a businesswoman working in a stressful, high-powered job never thought for a moment that something like a crystal could help her to relax. A friend gave her a piece of rose quartz; instinctively she responded to its color and started sleeping with it under her pillow. Suddenly her rest improved dramatically. Rose quartz is known to have extremely gentle healing effects on the whole system.

In the same way that different rays of color have different effects, if you are particularly drawn to, for example, green stones, it is likely that you need green energy in your life, perhaps some healing around your emotions. Choosing crystals based on color is a simple way to begin exploring the mineral kingdom and find out more about yourself and your energy needs.

BELOW *Water flows over rocks, the wind brushes trees. Everything is made up of particles with different levels of energy frequency.*

The World of Crystals

This chapter explores the mineral kingdom of Earth, showing how the planet we live on is in a constant state of change, and how the dynamic and powerful creative cycles of the earth cause crystals to grow inside it. We will also look at different crystalline structures, as well as different groups of crystals, to find out about the processes that created them. This brings in elements of geology, the study of how the earth was formed, and mineralogy, the study of the elements that make up crystals and precious metals.

It is important to recognize the precious nature of these rare minerals and treat them with respect. It is easy to forget that the reason they can be obtained is because they have been taken out of the earth, sometimes by direct human intervention. Staying aware of the processes that created crystals in the first place helps you develop a respect for the crystals you decide to collect. Whether you simply have them in your space, wear them, or use them for healing purposes, they are still the gifts of the earth.

The dynamic forces of the earth

Earth is a planet subjected to dynamic forces that constantly change its structure, thanks to the intense heat at its molten core. This heat causes chemical reactions between minerals; these are natural, inorganic chemical substances and approximately 2,500 different types exist. The mineral kingdom provides the chemical building blocks for everything on the earth: rocks, plants, animals—and human beings.

A CHANGING EARTH

Scientists believe the earth to be around 4.6 billion years old. It formed when our solar system was born out of clouds of gas and the dust of other stars. Heavy elements, such as nickel and iron, sank to form the hot core. Lighter minerals, including silicon and gases such as oxygen—the ingredients of quartz—form part of a thicker, semimolten layer called the mantle. Solid minerals float on the surface of the earth and form the rocky crust. As heat from the core pushes molten minerals upward, cooler rocks sink, melt, and are pushed up to the surface again. Thus the earth constantly recycles itself.

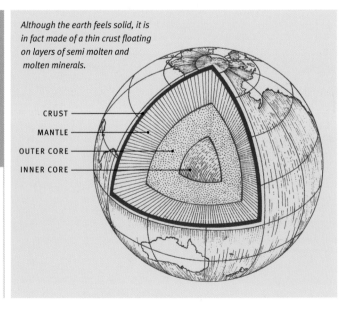

EARTH STRUCTURE

Although the earth feels solid, it is in fact made of a thin crust floating on layers of semi molten and molten minerals.

CRUST
MANTLE
OUTER CORE
INNER CORE

IGNEOUS • Igneous rocks can be formed either underground or above ground. Underground, when melted rock deep within the earth, called magma, becomes trapped in small pockets, it cools to become rock. Igneous rocks also form when volcanoes erupt, pushing molten lava up through fissures and cracks to the earth's surface; as this cools, crystals and rocks are formed.

When combined with water just below the surface, magma may form quartz crystals in large gaps called veins; depending on the minerals present, other gems, such as aquamarine, may also appear.

METAMORPHIC • These are rock layers that have been changed after their original formation, usually by increases in pressure, heat, water vapor, or chemical reactions. Layers of clay and sand sinking into the crust under pressure can form the mineral compound corundum, appearing as sapphire or ruby.

SEDIMENTARY • Sedimentary rocks are formed when deposits of material—plant, animal, or inanimate—are subjected to pressure, so squeezing out all the fluids and turning them into solid layers. The most common sedimentary rock is sandstone.

21

CONTINENTAL DRIFT

The earth's crust floats on the mantle. It is made up of different sections, called tectonic plates, and wherever these collide mountain ranges, such as the Rockies and the Himalayas, rise up; the fault lines between tectonic plates are also responsible for the creation of volcanoes and cause earthquakes. Deep in the oceans, cracks in the crust allow violent volcanic eruptions to take place, and these force the plates away from one another. Cycles of rising heated minerals and sinking cooler layers mean that pressure has to be released. New sources of precious stones and gems are always being found because the earth's surface is never still. Although formation takes millions of years, new mineral layers are continually being created (and destroyed) through the endless chemical reactions that are taking place within the earth's structure.

The structure of crystals

Most crystals are made up of highly structured and ordered patterns of molecules called lattices. They take on a stable and regular pattern repeated again and again through their internal structure, creating shapes and facets with particular types of symmetry. Some of them form within layers of rock, others inside large bubbles of gas so the crystals grow from the outside toward the center of the space. Factors such as pressure, temperature, and rate of cooling influence the shape a crystal will eventually take. The classification of typical crystal shapes is shown in the table below.

CRYSTAL LATTICES

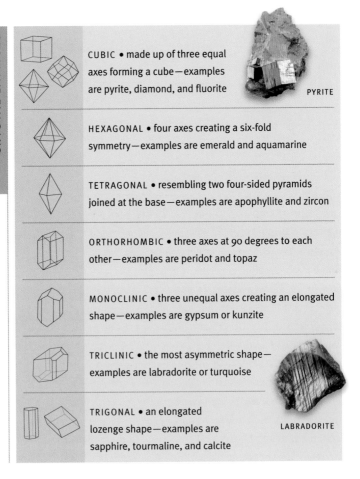

CUBIC • made up of three equal axes forming a cube—examples are pyrite, diamond, and fluorite

PYRITE

HEXAGONAL • four axes creating a six-fold symmetry—examples are emerald and aquamarine

TETRAGONAL • resembling two four-sided pyramids joined at the base—examples are apophyllite and zircon

ORTHORHOMBIC • three axes at 90 degrees to each other—examples are peridot and topaz

MONOCLINIC • three unequal axes creating an elongated shape—examples are gypsum or kunzite

TRICLINIC • the most asymmetric shape—examples are labradorite or turquoise

TRIGONAL • an elongated lozenge shape—examples are sapphire, tourmaline, and calcite

LABRADORITE

HARDNESS OF MINERALS

In 1812 a German mineralogist called Friedrich Mohs created a scale to help identify relative hardness in minerals. Hardness is a particular factor in jewelry making; only the hardest stones, such as corundum—ruby or sapphire—topaz, or diamond, are able to be faceted, that is, cut into shapes that allow them to reflect even more light so they sparkle. Softer stones shatter, which is why they tend to be simply smoothed and polished for setting in jewelry.

MOHS SCALE	HARDNESS	MINERAL	HARDNESS	MINERAL
	1	talc (softest)	6	moonstone
	2	gypsum	7	quartz
	3	calcite	8	topaz
	4	fluorite	9	ruby
	5	apatite	10	diamond (hardest)

OTHER FACTORS

SNOWFLAKE OBSIDIAN

Another classification, amorphous, is applied to specimens of organic origin such as amber, a fossilized tree resin, which does not have an internal geometric structure. Another type of amorphous stone is volcanic glass, called obsidian, which is formed when lava cools rapidly. Crystals are also assessed for their color, their luster—an inner quality, making them look waxy, oily, vitreous like glass, or pearly, for example—and their transparency. Another factor is called cleavage, which is the flat plane along which the stone splits naturally. Knowing this is vital in achieving the best results when shaping a stone.

Because crystals take form in an astonishing variety of shapes and sizes, it is rare to find perfect examples of the geometric shapes on page 22, but the more you examine the stones you buy, the more you will recognize some of these characteristics.

Main crystal groups

In mineralogy, the many types of crystals are classified into groups according to their chemical composition. The main groups shown on this page cover the kinds of crystals and minerals most commonly collected by crystal enthusiasts. In crystal healing, the quartz group is by far the most popular choice for treatment work. Quartz is able to act as a cleansing and amplifying energetic tool. Many healers use quartz wands to help rebalance their client's energy field or clear energetic blockages in specific problem areas. Quartz is found in almost every geological environment in the earth's crust.

CRYSTAL METHODS

From the earliest times, our ancestors gathered or dug up crystals from the surface of the earth quite spontaneously, and even today mineralogy enthusiasts still do. Mining for metals and crystals has taken place since the Stone Age, and in the Roman period there were well-established workings in the Middle East, supplying gems such as peridot. Modern industrialized mining for crystals takes place extensively in countries around the world, including Sri Lanka, Brazil, the USA, and South Africa. When you buy a crystal it is worth remembering it has had contact with the miner, the wholesaler, and the retailer—and maybe many others— before you purchase it. This is why cleansing crystals and preparing them for your own use is important (*see pages 282–83*).

RHODOCHROSITE

SPINEL

APATITE

ABOVE AND RIGHT *Crystals occur in a tremendous variety of color variations due to their mineral content. Some are commonly found and others are very rare.*

OXIDES • aluminum oxide forms the compound corundum, the red and blue forms of which are rubies and sapphires respectively; chrysoberyl, available mainly in golden yellow, is a beryllium aluminum oxide

NATIVE ELEMENTS • (pure metals): gold, silver, platinum, copper, carbon (as diamond), and titanium

SULFIDES • (metal plus sulfur): iron and sulfur forms pyrite, also called fool's gold

SILICATES • the main category is silicon dioxide, which makes quartz: clear quartzes—amethyst, rose quartz, smoky quartz, citrine, tourmalinated quartz, rutilated quartz; in microcrystalline quartzes, masses of tiny crystals appear in carnelian, agates, moss agate, and chalcedony, and these have a more waxy luster than clear quartz

also within this group are feldspars, comprising a huge number of crystals that are very common, including moonstone and labradorite

other large silicate groups are garnets, which may be rich in aluminum, iron, or chromium, depending on their location, and tourmaline and the beryls, including aquamarine or emerald, as well as the spodumenes, including kunzite

MINERALOIDS • tektites, such as moldavite, with a glasslike amorphous structure

ORGANIC MINERALS • organic, when talking about crystals, means a mineral that has plant or animal origins—amber, for example, is the fossilized resin of an ancient pine tree; pearls form when a grain of sand irritates the inner lining of the pearl oyster; and jet is a form of highly compressed coal

AMBER

25

Crystals in history

Humankind's fascination with crystals goes back to the roots of civilization. Prehistoric excavations of cave burials have demonstrated that ancient peoples prepared their dead for the next world by laying sacred objects on or around them, including different stones. It seems that these gifts from the earth have always held a fascination and been used to symbolize something special. Here are some examples of historically significant stones.

In China and the Far East, jade has been mined for thousands of years, and its deep green color is considered the most beautiful. The relative hardness of jade allows it to be carved, thus enabling craftspeople to shape the most exquisitely intricate statues, often of Chinese dragons. Priceless examples of jade statues were placed in the Chinese emperors' tombs as signs of their wealth and power.

Jade is also found in Central America, where it was sacred to Olmec, Mayan, and Toltec cultures. They carved jade into masks and sacred objects. The Aztecs considered emeralds to be sacred, and possessed examples large enough to be carved into goblets. The Aztecs were later invaded by Spanish conquistadores, who had heard of the wealth of precious stones and gold to be found.

JADE

LEFT *Ancient civilizations, such as the Olmecs of Mexico, carved masks of their warriors from large pieces of jade.*

OPPOSITE *The magnificent death mask of the boy pharaoh Tutankhamun is a masterpiece of ancient metalwork, inlaid with precious stones and gems.*

The Ancient Egyptians were skilled in the refining
and shaping of precious metals and stones. They
regarded deep blue lapis lazuli as sacred, and they used
it in combination with other stones, such as black onyx
and pure gold, to make exquisite collars, necklaces,
and other jewelry for the pharaohs, the higher nobility,
and priests. The pharaohs were also buried wearing
masks, such as the famous one of Tutankhamun, made
of gold inlaid with precious stones.

LAPIS LAZULI

In Europe, the Celtic people were noted for setting stones in
exquisitely worked gold. Flowing and curling patterns demonstrated the
Celtic love of nature. Intricate brooches or pins for cloaks in gold and
containing amber, garnets, or glass have been discovered—the beautiful
Tara brooch in the National Museum, Dublin, Ireland, is a fine example.

The medieval abbess and mystic Hildegard of Bingen, who lived in
southern Germany in the twelfth century, was also a writer on medical
topics. She advocated the use of many gems, including rubies and topaz,
as gem remedies to help with a variety of physical complaints. The gems
would be soaked in wine and the liquid taken as a medicine.

The Crystal Color Directory

This section of the book contains individual entries covering the crystals. In each entry, you will find geological data about the stone, its composition, and its most common sources, as well as its hardness and other physical characteristics. Traditional associations and uses of the crystals in healing is also covered. The chapter is subdivided into twenty sections relating to the twenty color bands shown on pages 12 and 13. The crystals have been placed in groups according to the shades of color in which they most often occur; any variations are also noted.

   As well as color associations, you will also find
links to history and/or crystal healing applications.
This information is given so that you can explore
and learn about any stone you buy from different
perspectives. Consider, interpret, and feel your way
into the descriptions; learn to trust your intuition.
Remember that human beings and crystals are made
up of different types of energy, but we live on the same
planet and we just take different forms. If a particular
type of stone attracts your attention, then it represents
something you need to rebalance your system.

THE BROWN RAY

The idea of the color brown may seem dull at first, but when this shade is explored we discover many different layers. Darkest brown, almost black, is one of the earth's colors, of the soil itself. Soil is the source of minerals and organic substances that provide sustenance to all living things, and the earth itself is the basis of everything that moves on its surface. Placing your bare feet on the ground is a very powerful way to bring your awareness into your physical body; you can tune into the ground beneath you and feel it like an anchor, strengthening and supporting you.

Imagining darker tones of brown can help you create your own inner vision of what our planet looks like inside. Fairy tales from Scotland and Ireland tell of a place deep inside the earth that is filled with its own special light.

Richer tones of brown are warming, sensual, reminiscent of clay, a form of soil that is supple and malleable. From the earliest times, human beings have put their hands into clay, molded and shaped it, and fired it to form beautiful objects, such as bowls and vases. If you want to experience this rich brown earth energy more practically, then learning to work with clay is a grounding and creative experience.

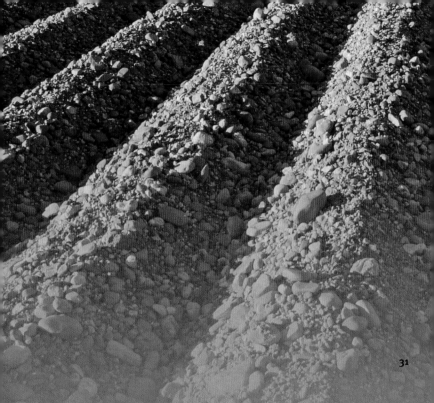

The healing earth

Many healers are now suggesting that there is an energy center in the human auric field—the energy matrix that surrounds the human body—that connects directly to the earth, with a deep brown color. If you stand on the ground, this energy center is considered to be about 12 inches (30 centimeters) under your feet. If you visualize it and breathe out deeply, you can send any feelings of negativity, anxiety, or stress down into the earth via this center, and, by inhaling deeply, you can draw the strength of the earth back into your system.

Crystals and stones in brown shades are generally seen as grounding, connecting with the earth, as well as being powerful personal cleansers and protectors of your environment. They are useful to carry around with you or to wear as jewelry if you are the kind of person who lives more "in the head" than "in the body," running around feeling that you are being pulled in all directions. The calming and grounding energies of brown stones will bring you back to a steadier sense of self.

SMOKY QUARTZ

Smoky quartz is a member of the extensive quartz (silicon dioxide) group of crystals with a distinctive brown shade. Some examples can be lighter and more transparent, while others are much darker and opaque; these have possibly been heat treated, so check with your supplier. Smoky quartz is clear quartz that has turned brown after exposure to natural underground radiation—for example, when found in granite deposits. For this reason, because it has already absorbed radioactivity, it is as if the stone "recognizes" other types of electromagnetic stress and neutralizes them.

Some examples of smoky quartz contain golden-colored strands of the mineral rutile, which is actually titanium ore, and this is called rutilated smoky quartz. The strands of titanium are said to act like conductors, speeding up the neutralizing effects of the crystal. Rutilated smoky quartz is a powerful space cleanser.

For healing purposes, smoky quartz points can be held or placed over areas of tension, or small polished stones can be held in both hands to release negativity. The energy of smoky quartz is both soothing and enveloping, encouraging you to rest and relax back into the strength of the earth. It is good to carry it on your person if you work in an environment filled with computers, or if you regularly spend time in a busy city with high levels of pollution or electromagnetic stress.

POLISHED

POINT

SMOKY QUARTZ

FORM AND STRUCTURE
trigonal, often in long, well-defined points or clusters

some examples are very large

COLOR
light brown to dark brown or almost black

GEOGRAPHICAL SOURCES
Brazil, Switzerland, USA

RARITY
easily obtained, both large pieces or small polished tumblestones

HARDNESS
7

PHYSICAL/EMOTIONAL USES
focuses and grounds a person's energy in the present moment, bringing peace and calm

helps ease pains in the lower back, hips, and legs

neutralizes negativity in the human energy field/aura

counteracts the effects of radiation—for example, electromagnetic fields around a television or computer

protects your home from negative energies

HEALING EFFECTS
in crystal-healing layouts, creates a safe and nurturing energy field for the release of negative emotions

place a large piece between the feet or smaller stones on either side of the body as protection

PERSONAL USES
place large pieces by electrical equipment, carry smaller stones with you, or wear as a pendant

hold a small stone in each hand to calm negative emotions

PETRIFIED WOOD

Millions of years ago this beautiful and unusual stone was originally part of a tree. Petrified wood is a fossil, created when forests were buried under layers of sedimentary rock, and the plant cells were literally bathed in minerals that replaced the organic structures in the original wood. The end result still looks like a tree, or at least part of a tree—even the tree rings are still visible. However, the material has the weight and feel of stone, and this is because it now has a microcrystalline structure, often called agate, made of silicon dioxide.

One of the largest petrified forests in the world, covering 93 square miles (150 square kilometers) in area, is on the Greek island of Lesvos. Fossilized tree stumps still stand upright with root systems in the bedrock. Approximately twenty-five million years old, they have been identified as types of tree now native to Asia and the Far East, showing how different the European climate was in that period. Other petrified forests are found in Canada, Australia, the USA, and Argentina, some of which are around 200 million years old.

In healing terms, petrified wood is old beyond imagining—it is a link to times so far in the past that we cannot picture them. With its mottled-brown appearance, it is a mediator between the organic and inorganic kingdoms: it was once a plant and is now immortalized in stone. It teaches a lesson of slow but inexorable transformation from one state to another.

SLICE

PETRIFIED WOOD

FORM AND STRUCTURE
wood fossilized by minerals in the microcrystalline quartz group, giving it an appearance and structure like agate

COLOR
mottled brown, black, reddish brown, depending on mineral combinations

GEOGRAPHICAL SOURCES
Argentina, Australia, Canada, Czech Republic, Greece, USA

RARITY
easily available in small polished pieces or large slices (which are sometimes used to make furniture)

HARDNESS
7

PHYSICAL/EMOTIONAL USES
cleansing to both the liver and the blood

aids stiff joints and arthritis

eases stuck emotions around old issues, clears old patterns linked to difficult relationships, especially of ancestral origin

is a symbol of slow evolution into new forms, so helps to connect spiritually to the deepest history of our planet

HEALING EFFECTS
in healing layouts, use on the pubic bone or over the base of the spine in order to stabilize the root chakra

PERSONAL USES
meditating with it can help to encourage the transformation of patterns that seem "set in stone"

BROWN JASPER

Jasper is a name given to a very large group of stones. They occur in many different colors due to different balances of minerals combined with microcrystalline quartz. They are often slightly dappled in appearance because of the tiny quartz crystals inside their structure, and show an astonishing variation of specks and stripes, depending on the particular combinations of minerals mixed with their basic structure of silicon dioxide.

Jasper can be carved, and it has been used to make personal seals, while larger pieces can be turned into vases or containers. In medieval times, jasper was a favorite with jewelers; stones were smoothed and rounded and set in gold to make rings, necklaces, goblets, or bejeweled plates for the wealthy and powerful. The twelfth-century German mystic and healer Hildegard of Bingen (*see also page 27*) used jasper stones to ease troubled dreams and improve sleep.

36

From a healing perspective, brown jasper helps build a connection with the earth. It grounds and stabilizes scattered energies. It is a gentle stone to use, and a good one to begin working with if you are new to crystal energy. Brown jasper also helps to clear and cleanse your personal space of negative influences. A beautifully striped brown jasper from Australia called Mookite—or Mookaite or Mook Jasper—supports the reproductive system and eases tension in the lower-back area; it also creates a connection with your ancestors, particularly if healing is needed around family issues.

POLISHED

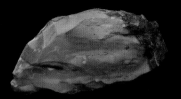

BROWN JASPER

FORM AND STRUCTURE
hexagonal microcrystalline quartz

impurities of limonite make the jasper brown

smooth texture

COLOR
brown, also red, yellow, green, purple, blue, multicolored; Mookite is reddish brown

all are opaque with many variations of specks and stripes

GEOGRAPHICAL SOURCES
Australia, Brazil, Germany, India, Russia, USA

RARITY
very common as larger pieces or small polished tumblestones

HARDNESS
6.5

PHYSICAL/EMOTIONAL USES
helps to clear toxins from the body and supports the immune system

strengthens the energy at the base of the spine

Mookite is used to help reproductive issues, and is a good stone to hold or carry during pregnancy

HEALING EFFECTS
brown jasper is similar to smoky quartz in that it neutralizes negative energies and electromagnetic interference

grounding and centering, enabling a connection with the earth

Mookite links to distant generations and helps dissolve old family issues

PERSONAL USES
place it on the forehead or hold it in the hands to calm stress and tension

meditate with it to stabilize your body within your energy field

THE RED RAY

Red is a deeply vibrant and powerful vibration. In color healing, it is used to improve the circulation and raise blood pressure. Red is challenging, strong, physical; it is the color of blood, the liquid that symbolizes our life force, the transporter of oxygen and minerals, nutrients, and hormones to every cell of the body. Red brings us into the arena of physical life, the physical body, and the business of survival. "Nature, red in tooth and claw" is a phrase that sums up this powerful energy.

Sometimes, in the pursuit of spiritual enlightenment, the basic, rooted, and powerful energy of red is pushed aside in favor of more "heady" and ethereal color vibrations. Red is seen as too primitive, too linked to our basic instincts. Yet these instincts keep us alive, and they are a powerful source of inner connection in themselves. Red is the blood of birth and the monthly blood of females, whose cycles are governed by the phases of the moon—the blood of creation. Creative processes can be very powerful, sometimes even painful, but they bring about a sense of incredible achievement.

The seven chakras

In healing, red is associated with an energy center called the root chakra. There are seven major chakra centers in the body, situated at the coccyx/ tailbone, the sacrum, the solar plexus area under the rib cage, in the center of the chest, the throat, the center of the forehead, and at the crown of the head. Each of these centers has a color, usually seen as following the rainbow sequence of colors (see pages 298–99 for more details). Red is the color that energizes the root chakra and enhances the physical life force in the body.

Red crystals, such as garnet or ruby, have a powerful effect on the physical body, often bringing a sense of warmth, stimulation, and comfort. In healing, they are used to generate more energy in the system if life force is low. Physical, mental, or environmental stresses can cause damage to the body, the precious vehicle we rely on to keep us alive, so making us feel cold, depleted, and lacking in energy when everyday demands take over. The rich luster and vibration of red stones renew our strength and our ability to take our lives into our own hands, make decisions, and take action.

GARNET

Garnets are actually a diverse group of stones taking many different forms and colors. They are formed in igneous or metamorphic rocks, and most of them are aluminum or calcium silicate minerals. There are six main geological types with many variations:

	TYPE	COLOR
	Pyrope	ruby red to dark red; most often used in jewelry
	Almandine	dark red to reddish brown; can be very large
GEOLOGICAL TYPES	Andradite	brown, black, or green
	Grossular	colorless, green, or orange
	Spessartine	pink, brown, or orange
	Uvarovite	green

The word garnet is derived from the Latin *granatum*, which means pomegranate, because the deep red color of garnets resembles the deep red of the seeds in the fruit. Garnets are sometimes considered inferior to more precious red gems, such as rubies, because they are cheaper and more available. This is a pity because garnets have a special beauty of their own. The Celts and early medieval jewelers made extensive use of garnets in rings, brooches, buckles, and necklaces. They were considered to be very protective.

In a healing context, garnets balance the mind, the emotions, and the body, bringing the heart into alignment with a higher expression of love. They help you discard old ways of thinking and old patterns that no longer serve. They open the way to abundance and vitality.

POLISHED

RAW

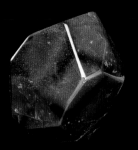

GARNET

FORM AND STRUCTURE
garnets can take many forms, the most common being a twelve-sided crystal with diamond-shaped faces

of the six main types, the pyrope or almandine varieties are most commonly associated with the name garnet

COLOR
pyrope garnets are dark to ruby red with a translucent shine when polished

almandine garnets are reddish brown

GEOGRAPHICAL SOURCES
pyrope garnets: Africa, Brazil, Czech Republic, Sri Lanka, USA

almandine garnets: Czech Republic, India, Sri Lanka, USA

RARITY
easily obtained, either as raw or gem-quality stones

HARDNESS
7

PHYSICAL/EMOTIONAL USES
pyrope garnets bring vitality and warmth to the system, improving circulation and promoting positive feelings of love

almandine garnets are more grounding, energizing the root chakra

HEALING EFFECTS
can be used in crystal-healing layouts to create a sense of "feet on Earth, head in Heaven," balancing mind and body

placed over the pubic bone, the heart, or between the feet they create a sense of true alignment

PERSONAL USES
wearing garnets as jewelry is the easiest way to experience their effects

traditionally the birthstone for the month of January

RUBY

Rubies are one of the most precious gems on the planet. They are the red variety of corundum, the second-hardest natural mineral after diamond. Chemically, it is an aluminum oxide, but traces of chromium make it red. Rubies can occur in large single examples or smaller clusters, and are found in deposits all over the world. Today, one of the best sources of deep red rubies is Myanmar, where they have a unique shade called "pigeon's blood." They are often set into exquisite jewelry alongside diamonds. Rough, uncut ruby is easily available and inexpensive; this is used in crystal healing.

Sometimes rubies contain strands of rutile (titanium oxide) that catch the light, causing a six-pointed star to appear. These are called "star rubies." Rutile is one of many imperfections that can change the shade or clarity of a stone. These imperfections distinguish a real stone from an artificial copy of one.

Rubies have been valued since the earliest times as symbols of royalty, set into crowns and diadems as an indication of wealth, strength, and power. The word ruby comes from the Latin *ruber*, meaning red. Rubies are the most frequently named gemstone in the Bible, and in the Old Testament, God places a ruby on Aaron's neck as the "lord of gems."

In healing, rubies can be used to make a very effective gem remedy (*see pages 284–85*) to increase life force in the body and arouse passion for life. Placing a ruby over the heart activates feelings of love and protects it from negativity.

POLISHED

RAW

RUBY

FORM AND STRUCTURE
trigonal structure, creating
distinctive hexagonal crystals
with a shining vitreous luster
when polished

COLOR
deep blood red, red to shades
of pink

GEOGRAPHICAL SOURCES
Afghanistan, Africa, India,
Myanmar, Pakistan, Sri Lanka,
Thailand, USA

RARITY
gemstone quality is readily
available and expensive, with
uncut stones easy to obtain
and inexpensive

HARDNESS
9

PHYSICAL/EMOTIONAL USES
strong affinity with the blood and
circulation, helping to warm cold
limbs and ease chilliness,
invigorating the system and
bringing renewed energy and
strength, and improving self-
confidence and courage

HEALING EFFECTS
in crystal-healing layouts, rubies
promote powerful and positive
feelings, helping to strengthen
the energy field/aura of the
person and encouraging spiritual
expansion

place them over the heart or the
pubic bone to increase life force
in the body

PERSONAL USES
make a gem remedy using
a raw stone

meditate with ruby to increase
vitality and independence

traditionally the birthstone
for the month of July

RED TIGER'S EYE

Tiger's eye is a name given to a group of stones with a special luminous quality in their structure. They are basically silicon dioxide—quartz—but they have formed in an unusual way, when fibers of a mineral called crocidolite are laid down in parallel bands within the quartz structure. This creates a silky-looking shimmer effect when the stone is turned to the light, resembling a cat's eye. The shimmering is sometimes called chatoyancy, after the French word for cat (*chat*). These stones are hard and can be carved; a wonderful way to appreciate the light effects in tiger's eye is to see a large piece in a polished sphere.

Whereas yellow tiger's eye is the most commonly found—along with a blue and a dark-banded variety called hawk's eye—red tiger's eye is also available. It is the yellow variety that has been heat treated to change its color. Heat treating of crystals is common, and there is debate among purists about the qualities of such stones. However, it could be argued that changes caused by heat happen inside the earth anyway, and in the case of red tiger's eye it does not change the structure. The reddish-brown shade of this stone has a beauty of its own and it is easy to find.

As a semiprecious stone, tiger's eye has been used extensively in jewelry since medieval times. It is usually cut and polished into a "cabochon," a smooth, oval shape, which brings out the chatoyancy in the stone and makes it a favorite for men's signet rings.

RAW

RED TIGER'S EYE

FORM AND STRUCTURE
hexagonal, easily shaped into
rounded, polished stones that
have a silky feel and reflect
the light

COLOR
red (heat treated), also yellow,
blue, blue-green

GEOGRAPHICAL SOURCES
Australia, India, Myanmar, South
Africa, USA

RARITY
easily obtained as small polished
tumblestones

HARDNESS
7

PHYSICAL/EMOTIONAL USES
red tiger's eye is warming and
grounding, helping to ease
sluggishness and "winter blues"

gently energizes the body and
awakens sexual energy

is a good stone to place in the
bath along with rose quartz to
soothe the emotions

HEALING EFFECTS
in healing layouts, red tiger's eye
can be used to energize the lower
abdomen, particularly the area
around the navel, soothing and
releasing emotional tension

PERSONAL USES
place on the pubic bone and
lower abdomen to ease stress

hold small tumblestones in
either hand to energize the
corresponding side of the body

RED JASPER

Jaspers are a varied group of stones. They occur in many different colors due to different balances of minerals combined with microcrystalline quartz. They are often slightly dappled in appearance because of the tiny quartz crystals inside their structure, with an astonishing variation of specks and stripes, depending on the particular combinations of minerals mixed with their basic structure of silicon dioxide.

Red jasper is colored by hematite (iron oxide) and it is found in massive natural formations. The color is really emphasized when it is polished. As far back as Roman times, jasper was used in exterior mosaics as well as stone cladding for the insides of buildings; it was also combined with marble to create magnificent floors. In St. Petersburg, Russia, there are a number of churches with entire columns carved from jasper in different shades, including red, black, and gray. It is a relatively hard stone, being a member of the quartz family, so it can be carved into any number of shapes. It is a favorite for creating personal seals and seal rings carrying a crest or motif.

Healing uses of red jasper concentrate on the raising of energies, increasing the ability to cope with all of life's pressures. It is seen as a strengthening and life-enhancing stone. It also enhances the root chakra, creating a connection with the earth. It is very soothing used in the bath along with other gentle stones, such as green aventurine, to support the system if energy is low.

RAW

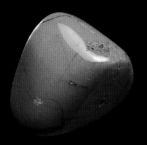

RED JASPER

FORM AND STRUCTURE
microcrystalline quartz in a
trigonal structure

opaque appearance with a
dense color

COLOR
red, also yellow, brown

GEOGRAPHICAL SOURCES
Brazil, France, Germany, Russia,
USA

RARITY
easily obtained and inexpensive
in large pieces, carved globes, or
small tumblestones

HARDNESS
6.5

PHYSICAL/EMOTIONAL USES
is gently energizing and
supportive, grounding and
protecting the body and
strengthening the energy field

energizes the circulation and
warms the system

promotes self-belief and the
courage to act

HEALING EFFECTS
in healing layouts, it can be
used on the root chakra or laid
in a circle around the body to
strengthen the energy field,
especially after a period of illness

PERSONAL USES
place under the pillow to
facilitate lucid dreaming

hold a tumblestone in each hand
to stabilize and balance energy

ZINCITE

Zincite is zinc oxide, a mineral that occurs in clusters of hexagonal shape. In its natural form, it is usually embedded in another base mineral such as calcite. Natural zincite is only found in a few locations. Other examples of zincite are not naturally occurring, meaning they are not mined or extracted from rock, but have actually formed as a result of smelting processes used in zinc mines. A notable example of this was seen recently in Poland, where zincite crystals of various sizes were discovered inside a smelting plant. They took form in beautiful colors, mostly shades of deep red or orange but occasionally green. These days most available zincite is being produced and sold as a by-product of zinc smelting. It has a structure and appearance similar to melted glass.

The astonishing colors of smelted zincite have attracted attention among crystal healers. The formation of these crystals through a reaction between intense heat and base minerals has a hint of alchemy about it. Alchemy was an ancient mystical science where base metals were processed to try to turn them into other substances, particularly gold.

In healing, zincite is seen as a potent symbol of transformation from one state to another, like the process of metamorphosis, by which a caterpillar turns into a butterfly. Zincite is used as a powerful stimulant of sexual energies, creativity, and fertility.

RAW

ZINCITE

FORM AND STRUCTURE
hexagonal in the natural state; as by-products of zinc smelting, glasslike with a translucent quality

COLOR
red, red-brown, dark red, orange, green

GEOGRAPHICAL SOURCES
Australia, Italy, Namibia, Poland, USA

RARITY
natural zincite is rare; the smelted variety is easier to obtain

HARDNESS
4

PHYSICAL/EMOTIONAL USES
used to raise the energy of the root chakra throughout the spine and the whole system, causing a profound increase in sexual energy and power (this is the foundation of all creative impulses—whether expressed through a relationship with another person or through individual creative pursuits—and in India it is called the rising of the kundalini serpent)

HEALING EFFECTS
place it over the base of the spine, the pubic bone or heart

best used in short applications; it is powerful and needs to be used with care because it has pronounced effects and the system needs time to get used to these effects

PERSONAL USES
a piece placed in your space is a powerful symbol of personal transformation

49

SPINEL

Spinel is an attractive gemstone, typically red, although sometimes with a pinkish tinge. It is magnesium aluminum oxide, and like ruby, obtains its color from chromium. Until recent times, all red stones tended to be considered rubies; only more precise mineralogical testing can verify the difference between them. Ruby has a hardness of 9 and spinel of 8, and this variation helps identification.

A famous historical example of a misidentified ruby is the massive Black Prince's Ruby in the British State Crown. This is, in fact, a large spinel of a beautiful red color, weighing $1\frac{1}{3}$ ounces (38 grams). It was obtained by the Black Prince in the fourteenth century as "payment" for his military services in Spain. A Moorish Prince of Granada had originally owned it. The stone was then set in the helmet Henry V wore at the battle of Agincourt in 1415, and in 1485, the doomed Richard III wore it at the Battle of Bosworth Field. In the early seventeenth century, James I had the stone set into the State Crown. Then, under Oliver Cromwell, the Crown Jewels were disassembled and the metal melted down. When the monarchy was finally restored in England in 1660, the spinel was miraculously recovered and incorporated once more into the State Crown.

In healing, spinel is used to create energy for life, new inspiration, and a sense of new beginnings. Another powerful stone, it anchors and opens the root chakra, paving the way for physical and spiritual expansion.

SMALL CRYSTAL

RAW

SPINEL

FORM AND STRUCTURE
cubic with pyramid-shaped
points, often in small spiny
formations, hence its name,
with a glasslike transparency

larger examples are found in
metamorphic bedrock, such
as marble

COLOR
typically red, also green, blue,
brown, purple, black

GEOGRAPHICAL SOURCES
Afghanistan, Brazil, India,
Myanmar, Sri Lanka

RARITY
available from specialist
suppliers

HARDNESS
8

PHYSICAL/EMOTIONAL USES
has an energizing effect on body
and mind, creating a positive
outlook and a refreshed approach
to the future

is used to detoxify the body
and is also capable of healing
emotional toxicity

HEALING EFFECTS
in grounding and cleansing
healing layouts, it is used on the
base of the spine, over the pubic
bone, or between the feet as an
"anchor" to prepare the body for
spiritual expansion

PERSONAL USES
place a piece in your workspace
to encourage inspiration

THE ORANGE RAY

From soft apricot shades to the vibrant color of the ripe fruit carrying its name, orange is one of the most cheering of all colors. It immediately conjures up images of summer vacations, spending time outdoors, or perhaps the taste and smell of the fruit on the tongue. Advertisements for orange juice frequently show happy people in the sunshine because the color and the fruit are associated with light and freedom. Children enjoy the color and taste of orange, and many adults need more orange energy to encourage a childlike sense of playfulness and spontaneity.

Orange is associated with the sacral chakra, which sits at the triangular sacrum bone at the base of the spine. This chakra energy is warming, stimulating, and antidepressant; it is also said to tone the reproductive organs. Orange energy is a combination of red (root chakra, connection with the earth, physical power) and yellow (solar plexus chakra, mental expansion and concentration). It builds confidence and the power of true expression within relationships, whether with the self or with others. Orange is needed when life seems dreary, when you need a pick-me-up, when you are stuck in fear, or feeling the effects of the middle of winter. Bright orange shades can be too stimulating, but softer, more muted apricot- or peach-colored shades are alternatives. They energize the sacral chakra energy more gently. Intuitive color healers will mute the orange ray according to their patient's needs.

The color of the sun

In some parts of the world orange and orange-gold colors have a special significance. In some Asian countries—India, Sri Lanka, and Thailand, for example—monks wear robes dyed in these shades because they are the color of falling leaves and, therefore, a symbol of letting go and nonattachment. Wreaths of orange or gold flowers are draped over statues of the gods; these are associated with the sun in all its changing moods, from the glow of sunrise or sunset to the brighter light of midday. In northern-European herbal medicine, the orange-colored flower called marigold, or calendula, was a favorite with early herbalists, such as the Englishman John Gerard. In the sixteenth century, he associated it with the sun and an ability to strengthen the spirits.

Orange crystals are a popular choice for jewelry because they are cheering and warming shades, enlivening any colors. In healing, these colored stones are used to help heal emotional imbalance and enhance self-confidence in emotional expression.

ORANGE CALCITE

Calcite is a common mineral accounting for about 4 percent of the earth's entire crust. It is calcium carbonate, most commonly known as limestone. In metamorphic rocks under pressure, it becomes marble; in caverns and caves, it forms stalactites and stalagmites. It is also the main ingredient in the shells of sea creatures. As they die, the calcium creates layers of sediment under the sea, which is the main constituent of limestone. Calcium carbonate crystallizes as calcite into a variety of forms, shapes, and colors, and often as large crystals in twinned pairs. It is available in an astonishing array of shapes and colors.

Generally, all the colored varieties of calcites are seen as energy transformers, dissolving the old in order to make way for the new. Orange calcite, with its soft shade and link to the orange color ray, is a gentle yet sustaining support to the energies of the sacral chakra. This stone works positively on those who give to everyone else and forget to give to themselves, particularly new mothers trying to get used to a completely new routine and the constant demands of a newborn child. Keeping orange calcite in the home, to look at and to hold sometimes, is a wonderful tonic to the spirits, for both mother and baby.

POLISHED

RAW

Orange

ORANGE CALCITE

FORM AND STRUCTURE
an amazing variety of shapes and
forms exist, and key forms are
rhombohedral, like large pointed
teeth, often in pairs

COLOR
orange, also clear, blue, green,
honey, pink, brown, black, gray

GEOGRAPHICAL SOURCES
Mexico

RARITY
easy to find in all colors, either
as large polished spheres, or as
smaller chunks or points

HARDNESS
3

PHYSICAL/EMOTIONAL USES
all the calcites have a gentle,
soothing look and feel, and the
orange variety is connected with
the sacral chakra, releasing
negative emotions and enhancing
zest for life

useful for enhancing creativity
and playfulness, which help solve
seemingly impossible problems

is employed to balance the
digestive system

HEALING EFFECTS
in the bath, use with other
soothing stones such as honey
or pink calcite to calm stress
and anxiety

in healing layouts, place over
the solar plexus or lower
abdomen to calm the digestion
or to prepare someone gently
for expansion into new levels
of energy

PERSONAL USES
hold one stone in each hand for
a soothing inner balance or as a
focus for meditation

ARAGONITE

Aragonite is another calcium carbonate mineral with the same chemistry as calcite but a different structure and shape. It is also harder and denser than calcite, with a more glasslike luster and amazing clusters of crystals in three-dimensional spiky formations, called sputniks after the Soviet spacecraft. Another common form is called flos ferri, meaning iron flowers, which is where aragonite grows in twisted, curving formations resembling coral. Its name is linked with the Spanish province of Aragon, where it was first found, although it is now also being sourced from Morocco, Britain, and the USA. In tiny amounts, aragonite as a mineral is responsible for the iridescent and pearl-like sheen inside the shells of such creatures as abalone or paua. Being another form of calcium carbonate, it is a major constituent in the bony structures of certain shellfish and corals.

The most commonly used crystals in healing are reddish-orange in color, from Spain or Morocco, although other colors do exist, including green, blue, or gray. The astonishing shapes and clusters and twisted forms of aragonite have an almost otherworldly look about them, like movement frozen in time. The complexity of the crystal formations readily suggests shapes and symbols to the mind, sometimes relating to past experiences or even past lives. This crystal is able to transmute energy from old experiences and pave the way for the new ones. The starlike structures help to create a connection to higher levels of awareness beyond the physical body and self.

CLUSTER

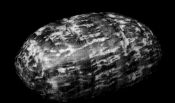

ARAGONITE

FORM AND STRUCTURE
orthorhombic formation with a
glasslike shine and texture, and
with many crystals radiating from
a central point or clustering
around a single central pillar

COLOR
orange-red, also brown, yellow,
green, blue, gray

GEOGRAPHICAL SOURCES
Britain, Germany, Morocco,
Namibia, Spain, USA

RARITY
obtain from specialist outlets

HARDNESS
3.5–4

PHYSICAL/EMOTIONAL USES
helps balance the emotions and
the physical body, easing feelings
that have been trapped inside

changes feelings of power-
lessness into self-confidence

helps support the skeleton and
the bones, the physical frame
of the body

its orange-red color energizes
the sacral chakra and warms the
body by gently increasing
circulation

HEALING EFFECTS
in healing layouts, used over the
lower abdomen aragonite clears
and heals negativity and
emotional blocks

over the heart or the throat it
facilitates feelings and
expressions linked to past events

PERSONAL USES
wear aragonite in a pendant over
the heart to ease the emotions,
or place it in a healing room to
enhance higher energy
frequencies

CARNELIAN

Carnelian is one of the most commonly available microcrystalline quartzes. Known as agate or chalcedony, it is made up of silicon dioxide colored by different levels of iron impurities, with specks, banding, or stripes in many shades of brown or orange-red. The stone itself has a beautiful warm orange color that is best appreciated when it is polished. It is available in a variety of shapes and sizes, as natural tumblestones or larger carved spheres or eggs.

Since ancient times, carnelian has been polished and worn as jewelry. The Egyptians used it to contrast with onyx and lapis lazuli in the making of collars and necklaces. The Romans were fond of it set in gold, using small beads in earrings or larger polished stones in finger rings for men and women. Because carnelian is a hard stone, it can be carved; examples of Roman signet rings or cameo rings survive from the early second century CE with exquisitely fashioned personal emblems or figures of gods. In the Middle Ages, carnelian was also popular as a healing stone. It was said to dissolve anger or rage, protect the wearer from negative influences, and promote courage.

Many people choose carnelian in the early stages of creating a crystal collection. Its warm orange color is attractive and it feels silky smooth when it is polished. It can be found as carved "worry stones" that can be carried and touched to ease stress and tension. Simple tumblestones in a wonderful variety of shades of orange make a lovely display.

RAW

CARNELIAN

FORM AND STRUCTURE
trigonal with fibrous layers of
quartz, creating soft smooth
bands of color and a hint of
translucence

COLOR
orange in different shades, from
pale to deep orange-red; stones
can be clear or show a wide
variety of specks, stripes, or
markings

GEOGRAPHICAL SOURCES
Brazil, India, Iran, Saudi Arabia,
Uruguay

RARITY
very easily obtained

HARDNESS
7

PHYSICAL/EMOTIONAL USES
speeds up healing processes,
especially after trauma or injury

helps improve elimination and
increases vitality and strength

purifies the blood and improves
circulation

helps bring courage when facing
personal challenges

HEALING EFFECTS
used with rose quartz over the
heart, carnelian balances sexual
energy with the vibration of love

in healing layouts, placed on the
lower abdomen, it energizes the
sacral and root chakras

PERSONAL USES
lay on the lower abdomen or
lower back, place in the bath,
or wear over the heart for warmth
and support

SUNSTONE

Sunstone is a member of an extensive group of minerals called feldspars. It has a complicated chemical structure, being sodium calcium aluminum silicate. As a mineral this stone is also known as oligoclase. Two types of oligoclase are sunstone and moonstone, both of which have light-reflecting properties. In sunstone's case, these arise out of small particles of hematite (iron oxide) held within the structure, giving a sparkling golden orange effect. Top-quality stones are extremely light reflective with a sparkling appearance; lower-quality examples will be pale colored with only a few visible shining particles. Sunstone is best polished in a round or oval shape to set off its light properties in jewelry, and small orange tumblestones are easily obtained from crystal suppliers.

Norway has been an important source of sunstone for hundreds of years. Translation of Norse texts has led to speculation that the Vikings, who were famous seafarers, may have used sunstone to catch rays of light from the sun to help them navigate when landmarks were out of sight.

The sparkling nature of sunstone makes it an attractive and refreshing stone to have around; it has an almost effervescent quality. It has a positive effect on the mind and the emotions, inspiring faith and increasing spiritual awakening. If you love the sun, then this stone will connect you to solar energy and fill you with light. It is good to have it around in the dark depths of winter.

RAW

POLISHED

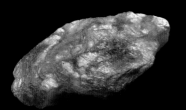

FORM AND STRUCTURE
triclinic system with a translucent
or milky appearance

forms in blocks or slightly slanted
termination points

COLOR
good specimens are orange with
plenty of inclusions; poorer
specimens are creamy white or
pale orange with few specks

GEOGRAPHICAL SOURCES
Canada, India, Norway, Russia,
Sweden, USA

RARITY
easily sourced from crystal
suppliers

HARDNESS
6–6.5

PHYSICAL/EMOTIONAL USES
helps lift depression and anxiety,
clearing the mind of negative
thoughts to enable positive
decision making

creates a feeling of expansion

balances the orange sacral
chakra with the yellow solar
plexus chakra, helping the mind
and emotions to work together

eases aches and pains and
rheumatism

HEALING EFFECTS
in healing layouts, place on the
lower abdomen to work on the
sacral chakra, or lay on the throat
to ease soreness or difficulties
with communication

PERSONAL USES
wear over the heart to spread
light and positive energy to the
whole body

meditate with the stone to
experience its energy

COPPER

Pure copper is one of the best conductors of heat and electricity. It is used in a wide variety of situations, including electronics, engineering, household products, plumbing, coinage, musical instruments, and glassmaking. Most reserves of naturally occurring native copper have now been exhausted, and today supplies are extracted from minerals such as malachite that contain copper as inclusions.

Human beings have used copper for at least ten thousand years. Archeology has unearthed remains of early civilizations in Anatolia, which shows that the smelting of copper occurred as far back as five thousand years ago. Evidence of early copper workings have also been found in China and the Andes (from four thousand years ago), Central America and West Africa (from around a thousand years ago). Copper is soft and easy to extract from other minerals using heat or by mining. It was used to make domestic items as well as spearheads and other weapons. Combining tin with copper creates bronze, an alloy that gave its name to an entire era in human history, the Bronze Age. In Roman times, the metal became known as *aes cyprium* (*aes* meaning alloy and *cyprium* meaning Cyprus, where there were mines). This was then modified to *cuprum* in Latin (the source of the chemical symbol Cu), becoming copper in English.

In healing, copper is used in the manufacture of some crystal wands, where one type of crystal, such as clear quartz, is combined with different stones to create a powerful working tool. Copper is chosen because it is a conductor of energy, assisting the harmonious working of the combination of stones in the wand. These types of wands are directional energy implements and should only be used by crystal-healing professionals who understand their effects.

NUGGET

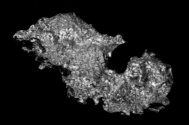

COPPER

FORM AND STRUCTURE
cubic metal forming natural smooth nuggets, or inclusions in other minerals such as malachite, azurite, and cuprite

COLOR
dull orange; green when oxidized

GEOGRAPHICAL SOURCES
Australia, Germany, Russia, USA

RARITY
natural-formed nuggets are occasionally available as collectable mineral samples

HARDNESS
3

PHYSICAL/EMOTIONAL USES
commonly worn as a bracelet in order to help ease rheumatic or arthritic pain

its properties as a conductor make it useful if physical or mental energies feel blocked or stuck

HEALING USES
in healing layouts, place copper over the thymus gland on the breastbone to support the immune system and build energy in the body

PERSONAL USES
display in arrangements with malachite to cleanse and harmonize personal space

THE YELLOW AND GOLD RAY

Bright as sunlight, energizing and expansive, the color yellow symbolizes the full strength of the midday sun at its height. It is a positive ray, stimulating mental clarity and keen observation. It helps with decision making when you need to "see the whole picture." Fresh tones of yellow are instantly brightening and cheering in the home, popular for interior decoration.

Yellow has a cleansing and detoxifying feel about it, clearing away debris to leave a clear space. Yellow fruit, such as lemons and grapefruit, have the same kind of appeal—bright, fresh, reenergizing. They are excellent in the diet to tone the digestive system and help the liver to function well; they also support the immune system, which benefits from the yellow ray in color healing.

This is the color associated with the solar plexus chakra center, situated in the middle of the upper abdomen where the ribs curve upward above the stomach. This chakra is linked to mental activity, expansion, and intellectual analysis, and it improves concentration. However, too much activity in this area can be limiting; the mind does not hold the whole picture when it comes to living one's life. The solar plexus chakra is frequently a site of tension, requiring balance from the heart chakra above, which brings love to the situation, and the sacral chakra below, which helps in relating to others.

Golden rays

Gold is a supercharged version of the color yellow. The metal gold is one of the most precious minerals available; it is also a symbol of the sun, not just the physical star in the sky but also the spiritual center of the cosmos. Gold is a beautiful healing ray, universal in its power to transmute negativity. If you are in need of emotional or spiritual comfort, imagine yourself in a shower of golden rays. Feel this transmuting energy. If you believe in angels, then the beauty of gold may speak to you of their presence, as it is traditionally the color of their haloes.

The metal gold and the yellow-colored crystals are a lovely group of minerals, highly attractive to wear and positive in their healing effects. They promote a state of well-being and joy. They reflect light and give a sense of clarity and ease. They promote self-confidence and self-expression in the highest sense. Collect and keep them in the home to increase cleansing and healing energies in your personal space.

GOLDEN TOPAZ

Golden topaz is a beautiful gemstone with a golden luster. Its chemistry is aluminum silicate fluoride hydroxide. It is hard, so both cutting and faceting bring out its brilliance; however, it splits in one direction in a way similar to diamond, so it needs to be handled carefully by jewelers. It sometimes occurs in large heavy specimens, but it is usually available to collect in smaller pieces, mainly eight-sided crystals, or octahedrons.

Possible derivations for the name include the Sanskrit *tapaz*, meaning fire. Topaz is also linked to an island in the Red Sea known as Topazion, reputedly a historical source of the gem, but, in fact, the stone most commonly mined there was peridot. The Ancient Egyptians considered topaz to be sacred to the sun god Ra, and the Romans linked it to their sun deity, Jupiter. It bestowed power and strength on the wearer.

Topaz has a long association with healing. In medieval times, it was believed to protect from the effects of magic, improve the eyesight, cure insomnia, cool boiling water, and promote spiritual inspiration. Many kings and nobles also wore topaz because it was said to change color in the presence of poison.

In healing practice, golden topaz is used to clear and strengthen personal will in line with higher intentions. We like to think we alone can decide what is right for us; however, sometimes our human perceptions are limited. Topaz opens us to higher inner wisdom and guidance.

RAW

GOLDEN TOPAZ

FORM AND STRUCTURE
orthorhombic system,
with glasslike transparency

fault lines often visible along
one plane

granular clusters can be large

COLOR
golden yellow, also clear, orange,
red, green, blue (usually heat-
treated clear topaz)

GEOGRAPHICAL SOURCES
Brazil, Mexico, Myanmar,
Pakistan, Russia, Sri Lanka, USA

RARITY
easily obtained as unpolished
specimens from crystal suppliers

set in jewelry

HARDNESS
8

PHYSICAL/EMOTIONAL USES
clarifies intent linked to higher
purpose, helping a person pursue
their true path

clears the head and improves
concentration

clears the solar plexus chakra
of negativity and supports
elimination via the kidneys

HEALING EFFECTS
in healing layouts, place over
the third-eye chakra between the
eyebrows to clear the mind and
open it to higher wisdom, or over
the solar plexus area to enhance
chakra energy and ease tension
in the abdomen

PERSONAL USES
use as a gem remedy (*see
pages 284–85*) to connect to
life purpose

wear as jewelry

hold a stone as a focus for
meditation

67

CITRINE

Citrine is a beautiful variety of quartz (silicon dioxide). Quartz is one of the most common minerals in the earth's crust, occurring in many different-sized specimens, shaped terminations or points, or clusters. Citrine is transparent with a pale gold color from traces of iron. This stone is rare in its true form, recognizable because of its defined points and pale hue. Darker yellow or brownish stones sold as citrine are, in fact, heat-treated amethyst; these are attractive but not correct for use as citrine in healing.

Citrine has been worn as jewelry for centuries—examples survive from Roman times—cut into rounded cabochon shapes and often set in rings. Today it is available in pendants, earrings, and rings, attractive to wear because of the soft golden luster. In addition, carved examples of citrine are available in spheres or flame shapes, carefully cut and polished to show light-reflecting faults or rainbow prisms in the crystal. The choice between naturally formed citrine specimens or artificially shaped stones is a personal one. As a suggestion, if you want to use stones for their healing properties, they are best used in their natural state. However, crystals shaped by skilled hands do have their own beauty.

In healing citrine is used to bring the energy field, or aura, of the person into alignment with the physical body. Stress and traumatic events can cause depletion, and the energizing qualities of citrine help restore balance. The sparkling beauty of the crystal also attracts abundance, so place it in the home to encourage a better flow of material energy.

POLISHED

CITRINE

FORM AND STRUCTURE
trigonal, in large defined points
or smaller clusters

COLOR
pale, golden yellow; darker
shades are heat-treated amethyst

GEOGRAPHICAL SOURCES
Canada, France, Madagascar,
Mexico, Russia, Spain, USA

RARITY
natural citrine is rare; consult a
specialist supplier

HARDNESS
7

PHYSICAL/EMOTIONAL USES
stimulates creative thinking and
inspiration, freeing the mind of
limitations

helps turn ideas into reality

supports hormone balance in
the body and helps to enhance
physical vitality

lifts depression

HEALING EFFECTS
in healing layouts, place on the
forehead to increase creative
inspiration or over the solar
plexus area to clear tension
from the upper abdomen

PERSONAL USES
place in your personal space
to bring good fortune and
abundance

meditate with a piece of natural
citrine for mental clarity

take a liquid citrine gem remedy
(*see pages 284–85*) to revitalize
the system

carry a crystal with you for
inspiration

AMBER

Amber is not actually a crystal or a gem, it is a natural fossilized resin from extinct evergreen trees, which has taken between twenty-five and fifty million years to harden. The most well-known source is the Baltic coast. Samples of amber have been found containing whole insects, perfectly preserved, allowing scientists to study them in detail. Many pieces contain seeds or pollen grains that were originally absorbed into the resin when it was sticky, then became trapped there permanently. Amber is most commonly found in beautiful yellow or deep golden shades, although green, red, and brown examples are also available. Because amber is a resin, it is very soft as a material—it is also lightweight and smooth to touch.

Archeological evidence shows humankind has been collecting and wearing amber since the Stone Age. In Ancient Greek times, it was considered to be the juice or essence of the setting sun, charged with solar energy, and sacred to Helios, the god who was said to drive the sun across the sky in his chariot. In Homer's epic poem the *Odyssey*, amber is described as being given as a priceless gift. Amber was also a sign of wealth and status in Celtic society, set in silver or gold, and fine brooches have been found in burial sites.

In healing, the golden-yellow color of amber is warming and cheering to the spirits, bringing a sense of renewal and recharged energy levels. It is extremely popular to this day as jewelry, mainly in rings, pendants, and earrings.

POLISHED

RAW

AMBER

FORM/STRUCTURE
amorphous fossilized resin,
in small pieces to large lumps

lightweight with a smooth
texture, opaque when raw and
a deeper color when polished

COLOR
golden yellow, also, more rarely,
red, brown, green

GEOGRAPHICAL SOURCES
Britain, Poland, Russia, South
America

RARITY
easily sourced

HARDNESS
2

PHYSICAL/EMOTIONAL USES
restores the stomach, kidneys,
and spleen and supports the
immune system

strengthens the lower back,
pelvis, and reproductive organs

regenerates energy in the solar
plexus (yellow) and sacral
(orange) chakra centers, helping
to stabilize and ground emotions

stimulates a positive mental
attitude and improves self-
confidence

HEALING EFFECTS
in healing layouts, place amber
pieces over the throat, the solar
plexus, the navel, and on each
side of the pelvis to regenerate
and strengthen the body

PERSONAL USES
wear on the hands as rings or
in pendants over the heart to
balance the emotions

YELLOW TIGER'S EYE

Tiger's eye is a name given to a group of stones with a special luminous quality in their structure. They are basically silicon dioxide—quartz— but they have formed in an unusual way, when fibers of a mineral called crocidolite are laid down in parallel bands within the quartz structure. This creates a silky-looking shimmer effect when the stone is turned to the light, resembling a cat's eye. It is sometimes called chatoyancy, after the French word for cat. These stones are hard and can be carved; a wonderful way to appreciate the light effects in tiger's eye is to see a large piece in a polished sphere. Yellow tiger's eye is the most common form of the stone, with bands of gold to yellow-brown reflecting a lovely golden shimmering light.

Tiger's eye has been a popular semiprecious stone for thousands of years. Tombs in the ancient city of Ur in Mesopotamia (modern Iraq) dating back to 2500 BCE have yielded gold set with agates, such as tiger's eye and carnelian. At the time of Alexander the Great in the fourth century BCE, Greek goldsmiths used tiger's eye as one of an extensive range of precious and semiprecious stones set in gold jewelry, including necklaces, rings, and torques.

In healing, tiger's eye is used to shield the energy field, or aura, from negativity, as well as clearing tension and mental blocks from the solar plexus chakra in the center of the upper abdomen. It brings a soothing golden resonance to calm and restore body and mind.

RAW

YELLOW TIGER'S EYE

FORM AND STRUCTURE
trigonal, forming in a hexagonal pattern
opaque with shimmering layers of quartz and crocidolite fibers

COLOR
golden yellow, also blue, blue-green, red (heat treated)

GEOGRAPHICAL SOURCES
India, Myanmar, South Africa, USA

RARITY
very easily obtained

HARDNESS
7

PHYSICAL/EMOTIONAL USES
linked to the energy of the solar plexus chakra, clearing mental blocks and encouraging focus on current issues

supports a healthy metabolism and prevents physical energy from becoming depleted

reputed to keep the eyes healthy and increase clear vision

HEALING EFFECTS
in healing layouts, place over the solar plexus area or abdomen to ease tension and protect from external influences

use to "anchor" the lower part of the body as a platform for spiritual expansion

PERSONAL USES
hold a tumblestone in each hand and meditate on the golden quality of tiger's eye

make a gem remedy and take to support clear vision (*see pages 284–85*)

YELLOW JASPER

Jasper is a name given to a large group of semiprecious stones. They occur in many different colors due to different combinations of minerals combined with microcrystalline quartz. They are often slightly dappled in appearance because of the tiny quartz crystals inside their structure, and show an astonishing variation of specks and stripes, depending on the particular combinations of minerals mixed with their basic structure of silicon dioxide. Yellow jasper is an opaque stone with a smooth, silky feel when polished. It has a mustard-yellow color, which is caused by the mineral limonite, and sometimes contains tiny black specks.

Jasper's hardness allows it to be intricately carved, and traditionally it has been used to make personal seals. Larger pieces have been used to make vases, containers, or even chess sets. In medieval times, jasper was a favorite with jewelers; stones were smoothed and rounded and set in gold to make rings, necklaces, goblets, or bejeweled plates for the wealthy and powerful. The twelfth-century German mystic and healer Hildegard of Bingen recommended jasper stones to ease troubled dreams and improve sleep. Generally, all the jaspers were considered to have protective powers and were often carried as personal talismans.

Yellow jasper is used in healing to energize the solar plexus chakra and relieve tension in the abdomen. It has a protective effect if you are experiencing excessive demands from other people or feel overwhelmed by external influences. It strengthens the physical body, especially after a period of illness.

TUMBLESTONE

RAW

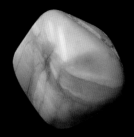

YELLOW JASPER

FORM AND STRUCTURE
trigonal system with a hexagonal pattern

microcrystalline quartz, colored yellow in this case because of the mineral limonite, with an opaque appearance

COLOR
yellow, also brown, red, black, green

GEOGRAPHICAL SOURCES
found worldwide, especially Brazil, France, Germany, Russia

RARITY
very common, available as small tumblestones or larger pieces

HARDNESS
6.5

PHYSICAL/EMOTIONAL USES
eases tension and stress in the abdomen, especially that caused by environmental effects or external pressures

improves the digestion and lifts emotional anxiety

cleanses the energy field/aura of negativity

HEALING EFFECTS
in healing layouts, place over the solar plexus chakra at the level of the diaphragm to ease tension and have a protective effect

place several stones in a circle around the body as a means of psychic protection

PERSONAL USES
carry yellow jasper to ground your energies and protect you from negative influences

place larger pieces in your home to cleanse and purify your space

SULFUR

Sulfur is a mineral available in several beautiful and unusual forms in its elemental state, and it is highly sought after by serious collectors. The name sulfur is from the Latin *sulfur* or *sulpur*, probably derived from the Arabic *sufra*, meaning yellow. This alludes to the bright lemon yellow of the crystals in their natural form. A classic association with sulfur—the smell of rotten eggs—is, in fact, sulfur dioxide, which occurs in organic sulfur compounds associated with plants or animals. Sulfur crystals have a faint aroma like matches. When they are burned, they emit a blue flame. This is the brimstone—an old name for sulfur—mentioned in the Bible. Sulfur crystals were used in alchemy, the mystical medieval branch of science that tried to turn base elements into gold. Sulfur, when mixed with potassium nitrate and carbon, is an ingredient in gunpowder, which was first developed by the Chinese in the tenth century.

Sulfur is also a key ingredient in plant, animal, and human biochemistry. It is essential to all living cells. In different chemical compounds, it performs a host of activities; as hydrogen sulfide, it is even a source of food for bacteria living in deep volcanic vents under the ocean. Sulfur compounds are absorbed from the soil by plants and the sulfur is used in several amino acids—the basic units of plant and animal proteins. Keratin, the protein in hair and fingernails, contains a lot of sulfur. It is the sulfur content in keratin that causes burned hair to smell peculiar.

In healing, sulfur is a powerful cleansing and purifying symbol, used to purify the system. Hot springs where sulfur combines with other minerals in water are excellent places to visit to detoxify the body.

MINERAL SPECIMEN

SULFUR

FORM AND STRUCTURE
orthorhombic, in well-shaped crystals or clusters, or pyramid faces

COLOR
strong yellow, also lemon or greenish yellow

GEOGRAPHICAL SOURCES
in hot springs or volcanic regions all over the world, especially Chile, Indonesia, Italy, Japan

RARITY
available from specialist suppliers

HARDNESS
2 (crumbles easily, handle with care)

PHYSICAL/EMOTIONAL USES
as a physical and emotional cleanser of the body and energy field/aura

to move through states of stagnation and negativity and encourage positive thinking

HEALING EFFECTS
in crystal-healing layouts, place between the feet to encourage stagnant energy to move down the body and out into the earth to be neutralized

PERSONAL USES
place in your environment to cleanse it of negativity

CHRYSOBERYL

This gemstone is not to be confused with true beryl, which is a silicate. Its chemistry is beryllium aluminum oxide, and it occurs in three main forms. The first—which was popular in Victorian and Edwardian times and can be found in antique necklaces, bracelets, earrings, and rings—is a beautiful yellow, yellowish brown, or yellow-green stone, hard enough to be faceted and set in formal jewelry. The second type is currently popular in the jewelry field. It is called cymophane or cat's eye; this is because when light strikes fibers in the stone they create a central sliver of light looking exactly like a cat's eye. This effect shows up particularly well if the stone is cut and polished into a rounded cabochon shape. Cat's eye chrysoberyl is usually honey yellow in color. The third and most expensive type of chrysoberyl is called alexandrite. It is very rare, and has a most unusual quality where it appears green in ultraviolet light and red in artificial light.

In India, the cat's eye chrysoberyl was said to protect the wearer from evil spirits, and Hindu philosophy linked it to increased health and prosperity. Placing a cat's eye between your eyebrows was said to increase your psychic abilities. The word chrysoberyl is partly derived from the Greek word *chrysos*, meaning golden.

Uncut samples of the golden yellow chrysoberyl are most commonly used in healing to help clear and energize the solar plexus chakra in the center of the rib cage.

CHRYSOBERYL FACETED

CAT'S EYE

CHRYSOBERYL

FORM AND STRUCTURE
orthorhombic

twin crystals that form rosette
structures are common

COLOR
common forms are yellow,
brownish yellow, or green;
cat's eye variety is honey yellow;
alexandrite is green in ultraviolet
light and red in artificial light

GEOGRAPHICAL SOURCES
Brazil, Myanmar, Russia (for
alexandrite), Sri Lanka

RARITY
not common, but available from
specialist suppliers

HARDNESS
8.5

PHYSICAL/EMOTIONAL USES
golden chrysoberyl is used
to align the personal will and
desires with a wider sense of
purpose, illustrated by the
phrase "not my will but thy will,"
attuning to a wider perspective
of "self"

physically helps to balance the
energy of the liver and gall
bladder

HEALING EFFECTS
in healing layouts, place on the
solar plexus chakra, over the
heart, or over the crown of the
head to facilitate the alignment
of desire and will

PERSONAL USES
wear over the heart in a pendant,
carry it with you, or meditate with
a stone to stay focused on your
true purpose

PYRITE

Pyrite is iron sulfide, and is sometimes known as fool's gold because of its metallic luster and brassy yellow hue. Tiny inclusions of pyrite are found in lapis lazuli. It is a heavy metallic mineral found in a huge variety of shapes and clusters. Pyrite has the same chemistry as marcasite but a different structure; marcasite tends to crumble into powder as it ages. (Confusingly, in jewelry circles, what is referred to as marcasite is actually pyrite—because the latter can be shaped and set.) Pyrite is found in veins of quartz as well as in sedimentary and metamorphic rocks, and even in beds of coal. The name pyrite is derived from the Ancient Greek *pyr*, meaning fire, so called because striking pyrite against steel makes sparks. Pyrite needs to be stored at low temperatures and to be kept dry because it can become brittle if exposed to humidity.

In healing, pyrite is used to build up masculine energy, the vital force that enables us to be active in the world, to "take life in our hands." In both men and women, a balance is needed between active, outward, and directed masculine energy and passive, inner, and reflective feminine energy. It is easy for physical vitality to become depleted, diminishing our confidence to shape our own destinies. Pyrite helps to reawaken our ideas and gives us the energy to make them a reality. Use hematite with it as a silvery-colored mineral companion to complete the balance of male and female energies.

RAW

PYRITE

FORM AND STRUCTURE
cubic, octahedral forms, often massive pieces or granular clusters, nodular and intergrown with each other

COLOR
brassy yellow

GEOGRAPHICAL SOURCES
Germany, Italy, Peru, Russia, South Africa, Spain, USA

RARITY
easily obtained

HARDNESS
6–6.5

PHYSICAL/EMOTIONAL USES
to clarify mental processes, chase away confusion, and shed light on problems

encourages decision making and taking action

supports the immune system

is reputed to help with male sexual or infertility issues

HEALING EFFECTS
in healing layouts, place over the solar plexus area in the middle of the ribs or hold a piece of pyrite in the right hand to balance masculine energy

if a person is lying face down, place pyrite over the kidney areas on either side of the back just under the rib cage to assist detoxification

PERSONAL USES
place pyrite in your office to help you stay focused, where it will also help to detoxify your environment

men can carry a piece to help enhance sexual vitality

GOLD

There is no more alluring and seductive mineral than gold; it has long been prized for its beauty, and has attracted and fascinated humankind for thousands of years. It has strange qualities, being very soft for a metal and yet virtually indestructible. It has been used and reused for centuries, so all the gold still in existence today is reckoned to be equal in volume to all the gold that has ever been mined. As one of the most precious of all metals, gold is still one of the most potent symbols of wealth and power.

Gold is a native element given the symbol Au from the Latin *aurum*. It is often found as streaks in white quartz, in veins called lodes, or in alloys with other minerals such as silver. Early samples of jewelry goldwork from the ancient city of Ur in Mesopotamia (modern Iraq) date back as far as 3000 BCE. Egyptian work from a similar period includes examples inlaid with precious gems. Amazingly intricate gold jewelry—drinking cups, vases, weapons, and ornaments—have been found in excavations at the ancient cities of Troy, Mycenae, and Tiryns, dating from between 700 and 500 BCE. In the Far East, India, Tibet, and China have been producing ornate and intricate goldwork for centuries; in Europe, Celtic jewelers used gold to make fabulous brooches and pins. In the sixteenth century, the Spanish invaded the Aztec and Inca civilizations in Central and South America primarily to find their legendary wealth in gold.

NUGGETS

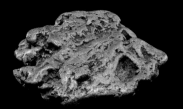

GOLD

FORM AND STRUCTURE
forms in nuggets, grains, or veins; is also found in other minerals such as quartz

COLOR
buttery yellow

GEOGRAPHICAL SOURCES
Canada, Russia, South Africa, USA

RARITY
as a mineral sample it is rare; it is more common to wear it as jewelry

HARDNESS
2.5–3 (it is very malleable)

PHYSICAL/EMOTIONAL USES
is associated with the sun, with solar energy and power

supports the immune system and brings strength and clarity to body and mind

balances and harmonizes the body's natural electrical field, especially useful in areas of intense environmental stress and radiation

HEALING EFFECTS
not often used in crystal-healing layouts because pure nuggets tend to be quite small, although visualizing light in a golden shade is one of the most generally soothing and restoring applications of color healing

PERSONAL USES
wear set with gems or crystals in order to enhance your self-confidence and self-esteem

THE GOLDEN GREEN RAY

Golden green blends the bright golden yellow of sunlight with the lush pale green of new leaves. Imagine sunlight shining through tender foliage in the springtime; this creates a gentle yet lush green color. Gold—linking to solar energy—is creative and regenerating in its effects. Green symbolizes qualities of growth and expansion. The combination of these two illustrates the way that light—the power of the sun—acts with structures called chloroplasts in plant cells to create the green color we associate most commonly with nature.

The quality of springtime is a very useful image to apply to this healing ray. Spring is symbolic of warmth and renewal, an awakening of new vitality after the cold, quiet months of winter. As buds burst open and spring-green leaves appear, feelings of optimism and cheerfulness develop. From a healing perspective, springtime is also a good opportunity to cleanse the body and mind and detoxify the system, particularly the liver. Eating fresh green spring vegetables is one way of doing this—taking that new plant energy directly into your system.

The color of renewal

In classical times, the color of spring green was celebrated in the form
of a goddess as a young girl, a free spirit at home in nature. The Greeks
called her Artemis, the maiden huntress; the Romans named her Diana.
Her youthfulness and energy are symbolic of this golden-green energy.
This shade also encourages expansion to new horizons, opening up new
ways of thinking and feeling, as well as growth in self-awareness. Golden
green can also be seen as a symbol of new love; wear this shade if you
want to attract that energy into your life.

We all need inspiration at times, especially if we feel stuck or unable
to move forward. Sometimes life can feel restrictive and we run out of
steam, especially when it comes to following our heart's desires. In color
and crystal healing, golden green brings renewed strength and zest to
the whole system, cleansing away negativity and refreshing the spirits.

Golden green also helps build a bridge between two major chakras—
the solar plexus (yellow) and the heart (green). The solar plexus is linked
to the mind, which often wants to control things, and the heart is far
more concerned with feelings. Sometimes these two elements can be in
conflict. Golden green enables balance, so that feelings can flow and the
mind can relax.

PERIDOT

Peridot is the gemstone variety of olivine, a name given to a mineral made up of a magnesium, iron, and silicate compound. Higher proportions of magnesium and iron are key features in identifying peridot; chromium and nickel impurities also contribute to its distinctive green color. It is often found in areas of volcanic activity—for example, in Hawaii, on the island of Oahu, pale green peridot grains are often washed up on beaches. Some specimens also appear in iron-nickel meteorites called pallisites.

As far back as three thousand years ago, the Egyptians mined peridot on an island in the Red Sea they called Zebirget, and the island remained an important source of the gem right up to the early 1920s. Although about 80 percent of all world production of peridot currently comes from Arizona, new sources of fine stones are now being found in Pakistan.

Peridot has been in use for over 4,000 years—it is said to have been a favorite with the Egyptian queen Cleopatra—and it was regarded as a powerful protective stone against evil spirits, particularly when set in gold. In medieval times, it was often used in the design of sacred spaces, such as the Shrine of the Three Kings in Cologne Cathedral in Germany.

In crystal healing, peridot is used to cleanse and detoxify the liver and gall bladder. It very effectively cleanses the heart chakra (green) of negative emotions and the solar plexus chakra (yellow) of overriding mental blocks.

POLISHED   FACETED

PERIDOT

FORM AND STRUCTURE
orthorhombic, opaque when raw, clear and lustrous when polished; mostly found as small pieces with a beadlike shape

COLOR
olive green or yellow-green stone, occasionally brown

GEOGRAPHICAL SOURCES
Brazil, Myanmar, Pakistan, Russia, Sri Lanka, USA

RARITY
easy to obtain, although more often as jewelry than raw pieces

HARDNESS
6.5–7 (can be brittle so is more easy to preserve when it is set in metal)

PHYSICAL/EMOTIONAL USES
neutralizes physical, emotional, or mental toxicity

encourages tissue regeneration and supports liver and gall bladder function

clears old life patterns to encourage new experiences

strengthens the heart and solar plexus chakras

encourages mental clarity and clear decision making

HEALING EFFECTS
in healing layouts, place raw stones over the heart or solar plexus chakras

place stones on the left and right sides of the body in order to cleanse the aura

PERSONAL USES
wear a peridot pendant over the heart, especially to attract love into your life

traditionally the birthstone for the month of August

CHROME DIOPSIDE

Diopside is an unusual mineral that occurs in deposits of metamorphic or igneous rocks and even in meteorites. It is composed of calcium magnesium silicate and is often found in clusters of short prismatic-shaped crystals with a glassy look. The light green variety is called chrome diopside, and is fairly easy to find. There are other types, such as a rare blue diopside from Italy called violan and a black type called star diopside. This contains microscopic needles of the mineral rutile, which gives a starlight effect called asterism when polished in a rounded cabochon shape.

Chrome diopside has a beautiful golden green color and is also dichroic, meaning it shows a darker color from certain angles. In crystal healing, it is linked to the golden green ray of new growth and renewal, bringing its energy into the entire system. Keeping chrome diopside nearby is helpful when recovering from physical trauma or a period of illness; it helps the body to regain its strength and encourages restoration at the cellular level. This crystal also helps create a deep connection with the earth, keeping the body grounded and centered; this is very helpful to people who are out of touch with their physical bodies and who live too much inside the head. Chrome diopside's color also establishes a symbolic connection with the realm of nature—the green, organic, living aspect of the earth.

BLADE

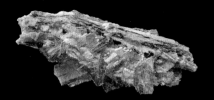

CHROME DIOPSIDE

FORM AND STRUCTURE
monoclinic system, forming short prismatic-shaped crystals, mostly in clusters but sometimes in massive or columnlike shapes with a glassy look and showing dichroism (two different colors)

COLOR
golden green, also white, blue, violet, black, yellow, or yellowish brown

GEOGRAPHICAL SOURCES
Austria, Finland, Germany, Italy, Russia, South Africa, USA

RARITY
all kinds of diopside need to be sourced from specialist suppliers

HARDNESS
5–6

PHYSICAL/EMOTIONAL USES
helps the body recover from physical or emotional trauma, transmitting a gentle golden green energy to help regenerate the system

restores vitality and optimism

realigns the body and the chakra system in line with the earth's electromagnetic field

releases stress on physical and emotional levels and enables deep relaxation

HEALING EFFECTS
in healing layouts, place on any part of the body that feels traumatized, along with a piece of smoky quartz between the feet to enable any negative energy to pass into the earth

PERSONAL USES
place in the home to feel more connected with the earth and the realm of nature

SERPENTINE

Serpentine is a beautiful, soft green-colored mineral mostly found in massive microcrystalline formations of metamorphic or igneous rocks. One type of serpentine is fibrous and a source of asbestos; however, microcrystalline serpentine deposits have a different chemistry and are safe to collect, and these are the specimens you will get from crystal suppliers. Serpentine forms in layers of silicates that create a smooth sheen and waxy texture when the stone is polished. Its chemical name is magnesium iron phyllosilicate.

Serpentine possibly gets its name because it has a mottled golden-green color similar to the skin of a snake—in fact, in ancient times it was said to heal snakebites. Plants do not grow well if there is serpentine in the soil or the bedrock because of its high levels of nickel, chromium, and cobalt; only conifers or tough types of shrub can survive in what are sometimes called serpentine barrens.

In classical times, serpentine was mined in Thessaly in Greece and transported to Rome. There, it was used to create the interiors of imposing buildings, polished and combined with other stones such as marble to create staircases, internal walls, or pillars.

Many types of serpentine around the world are still carved into statues or ornaments. For example, the Maori of New Zealand use the local variety to carve sacred objects, valuing its green color. It is found in Ireland, where it is called Connemara marble, and in Britain, where a particular form of it is called lizardite, after an area of Cornwall known as the Lizard.

POLISHED

RAW

SERPENTINE

FORM AND STRUCTURE
occurs in dense, massive forms; microcrystalline serpentine has an opaque appearance and a waxy feel when polished

COLOR
mainly soft golden green, mottled with specks of brown and black; yellow, brown, or black specimens are also found

GEOGRAPHICAL SOURCES
Britain, Canada, Ireland, Italy, Russia, Switzerland, USA

RARITY
not common; obtain from specialist suppliers

HARDNESS
3–4

PHYSICAL/EMOTIONAL USES
used to connect with one's own life purpose and clear away psychic debris from the past, whether in this lifetime or through ancestral links

is calming and soothing to states of emotional stress

HEALING EFFECTS
place over the heart chakra in the center of the chest to alleviate emotional pain or trauma

hold a piece in each hand to stabilize the energy field around the body

PERSONAL USES
meditate with serpentine to open yourself to greater awareness of your place in the world and your own path within it

THE PALE GREEN RAY

As well as occurring in different shades, all colors are also visible in different intensities. In healing, paler tones are often associated with more subtle or "higher" vibrations, linking to the unseen realm of the spirit. Imagine a note played on a piano, and then the same note played an octave higher. It is a similar case with color. Imagine a vibrant green shade, and now a much lighter one. Both are green, but the lighter one shines with a more subtle luminosity.

The pale green ray has a cool feel in vibratory terms. It is the color of ice over a pond in winter, where plants lie underneath, frozen and waiting for the spring; or like meltwater coming straight off a glacier in the mountains, supercharged with oxygen and, therefore, reflecting that strange opaque pale green color. This energy level is in direct contrast to the warmer colors we have seen so far. These cooler vibratory frequencies are necessary because they help to calm excess heat and excitement, bringing body and mind back to balance.

The faerie color

The soft, subtle shade of the pale green ray can also be seen as the gateway to what used to be called "faerie," that is the realm of fairies, elves, sprites, and goblins, elemental spirits and magic that exist in a world just beyond what we can see. In ancient belief, these small beings were imagined to be custodians of plants and flowers, sometimes tiny, sparkling creatures who danced in fairy rings, sometimes guardians of the trees and the stones or crystals that came from the earth. Working with this subtle level of pale green can open the doorway to magic, the unseen forces of transformation that work beyond the mind or logic. Pale green crystals have a delicate yet mysterious quality that helps to expand awareness into new realms.

If understanding the meaning of different levels of color energy is new to you, allow yourself to experience the more subtle shades simply by contrasting them with more vibrant tones. As you continue to make the comparison, ask yourself: How do they feel different? Why do they feel different? We all have the capacity to sense, feel, and interpret our world in many ways. Working with varied intensities of color allows your own individual sensitivity to grow and develop, as you learn to trust your intuition.

GREEN APOPHYLLITE

Apophyllite is a beautiful stone found in several colors, including a beautiful pale green shade with a pearly sheen on its straight facets. It tends to peel or flake apart if heated because its structure contains water molecules, so it needs to be kept cool if you collect it. Apophyllite is a generic term used to describe three different minerals with similar chemistry and physical characteristics. One type, specifically called fluorapophyllite, is the most common and the type most often sold with the label "apophyllite." Its chemical description is hydrated potassium calcium sodium silicate fluoride hydroxide. It formed in ancient lava flows or basalt deposits, inside bubbles created as the molten rock cooled. It is often found embedded in metamorphic rocks. Apophyllite is becoming more and more popular as a mineral specimen with collectors; in its green form, it tends to be more expensive than the more commonly available white clusters because it is rare.

In crystal healing, green apophyllite is used to open the doors of inner perception, expanding psychic and supersensory capacities. It is linked to the fairy or elemental kingdoms, as well as angelic levels of energy, enhancing a person's experience of these other realms. It brings a gentle and childlike energy to healing, encouraging lightness of body and mind and the release of heavy emotional burdens.

On a practical level, some commentators link apophyllite closely to the earth. They suggest placing larger crystals outdoors in a garden to enhance all the energies present in that space, so it becomes a haven for humans, birds, animals, plants, and the elementals, as well as a place of peace and harmony.

RAW

APOPHYLLITE

FORM AND STRUCTURE
tetragonal system, often forming in cubic masses of small faceted crystals or in pyramid shapes

sometimes occurs embedded in metamorphic rocks

COLOR
pale green, also clear, white, yellow, or violet

GEOGRAPHICAL SOURCES
Brazil, Germany, Greenland, India, Iceland, Mexico, USA

RARITY
green apophyllite is rare, but is available from specialist suppliers

HARDNESS
4.5–5

PHYSICAL/EMOTIONAL USES
can be used to heal childhood issues, releasing old emotional patterns and encouraging lightness and spontaneity in life

refreshes the mind and clears the heart chakra of emotional debris

HEALING EFFECTS
place over the heart chakra in healing layouts to encourage loving connection with nature in the widest sense, or to balance negative emotions

PERSONAL USES
try putting a piece in your garden or on a balcony to enhance outdoor space

meditate with apophyllite to increase your psychic awareness of other realms

PREHNITE

This is another beautiful, pale green mineral, which originates in the Eastern Cape, South Africa. Its name refers to the Dutch mineralogist Colonel Hendrik von Prehn, who brought the first specimens of prehnite to Europe in the eighteenth century. It is the first mineral ever to be named after a person.

Prehnite has a milky opaque sheen when polished and often forms in masses that are called botryoidal, which means grapelike, or sometimes as nodules or encrustations over other types of bedrock. Prehnite is found in deep caves or vents of metamorphic rocks in the earth's crust. It gives off water when heated. Its chemical description is calcium aluminum phyllosilicate.

In crystal healing, prehnite is valued as a calming and soothing stone, helping to ease emotional stress. Its pale green color is once again associated with spiritual levels of energy and expansion into higher levels of awareness. It is a gentle facilitator of inner transformation, promoting serenity and inner peace. It acts upon the heart but at a level beyond simple personal feelings, enhancing unconditional love for all beings. Prehnite is also a good stone for healers to use on themselves. When doing a lot of treatment work for others, it is easy to lose sight of the balance between giving and receiving, forgetting that everyone needs replenishment. Working with prehnite in meditation helps to heal the healer and restore energy that may have become depleted.

RAW

PREHNITE

FORM AND STRUCTURE
orthorhombic system in massive botryoidal, granular, or encrusted forms, with a waxy sheen when polished

COLOR
pale green, usually opaque, also yellow, white, colorless

GEOGRAPHICAL SOURCES
Australia, Britain, Canada, France, Germany, India, South Africa, USA

RARITY
easily available

HARDNESS
6–6.5

PHYSICAL/EMOTIONAL USES
helps to expand the heart into unconditional love, gently soothing and calming the nerves and the entire system, helping to release emotional pain from the past and encourage a new beginning

is a gentle detoxifying stone for the kidneys and lymphatic system

HEALING EFFECTS
use in healing layouts over the heart, throat, or third-eye chakra to expand the sense of self into wider awareness of creation

PERSONAL USES
meditate with prehnite to experience peace and calm

sleep with it under your pillow to encourage soothing dreams

GREEN CALCITE

Calcite is a very common mineral accounting for about 4 percent of the earth's entire crust. It is calcium carbonate, most commonly known as limestone. In metamorphic rocks under pressure, it becomes marble; in caverns and caves it forms stalactites and stalagmites. It is also the main ingredient in the shells of sea creatures; as they die, the calcium creates layers of sediment under the sea, forming over time into such rocks as chalk. Calcium carbonate crystallizes as calcite into a huge variety of forms, shapes, and colors, often as large crystals in twinned pairs. It is available in an extremely varied array of shapes and colors.

Generally, all varieties of calcites are seen as energy transformers, dissolving the old in order to make way for the new. They are one of the softest of all crystals—it is possible to scratch their surface with a fingernail. Green calcite has a lovely pale hue, soothing and gentle to the eye. It encourages a deeper relationship with the natural world, reconnecting us with the planet as a source of life energy. In crystal healing, this green crystal helps soothe unruly or fiery emotions. It can rebalance the heart chakra, encouraging compassion and understanding. Its energy is cooling and calming, bringing peace and inner balance.

RAW

GREEN CALCITE

FORM AND STRUCTURE
an amazing variety of shapes
and forms exist; key forms are
rhombohedral, like large pointed
teeth, often in pairs

COLOR
green, also clear, orange, honey,
pink, brown, black, gray

GEOGRAPHICAL SOURCES
found in fine examples in Mexico,
also in Brazil, Britain, Germany,
USA

RARITY
extremely easy to find, as large
polished spheres or as smaller
chunky pieces or points

HARDNESS
2.5–3

PHYSICAL/EMOTIONAL USES
all the calcites have a gentle,
soothing look and feel, and the
green variety is connected with
the heart chakra, transforming
anger or jealousy into
forgiveness and release

helps the body to release
emotional trauma and supports
the immune system

soothes jangled nerves and
calms tension or stress

HEALING EFFECTS
in healing layouts, place over
the heart chakra to facilitate
transformation of powerful
emotional states

place over the upper abdomen
to ease nervous tension or a
churning stomach

PERSONAL USES
hold one stone in each hand for
a soothing inner balance or as a
focus for meditation

place a piece of green calcite
under your pillow to increase
the clarity of your dreams

THE GREEN RAY

This is the richest, most lush shade of green, seen at its most beautiful in the vegetation of the tropics. The reason why plants are at their most green in tropical locations is because they have maximum access to the sun at its height, being closer to the equator. Provided they have enough water, the power of intense sunlight acting on plant tissues produces the most incredible emerald shades of green; this is the process of photosynthesis—the miracle of light creating food for plants—at its best.

This vibrant green is a symbol of growth, expansion, power, and vitality. It is no coincidence that the Aztec and Inca peoples of Central and South America valued green stones such as emerald and jade—these were visible symbols taken out of the earth that echoed the lush vegetation of their own surroundings. Green is also a powerful color to the Maori people of New Zealand, whose lands are filled with abundant plant life. Green as a color is used directly on the body in healing to help speed up regrowth of damaged tissues, and even to help heal broken bones. It helps to infuse body and mind with transformative energy, creating possibilities to move forward.

A bridge between body and soul

In the chakra system, green is the principal color associated with the heart. In the system of seven major chakras, the heart center is in the middle. Below it are the warmer vibrations of yellow (solar plexus), orange (sacrum), and red (base of spine); above the heart are the cooler vibrations of blue (throat), indigo (third eye), and violet (crown). Green is the bridge between these different energy levels; the energy of the heart chakra bridges the demands of the physical body, symbolized by the lower chakras, and the aspirations of the spirit, centered in the upper chakras. It does that by enabling the expression of love, which in its purest form balances all things.

Green is a good color to bring into your life when you feel a need for something new, when you want to break out of an old routine that no longer serves you, or when you are ready to start a new relationship. In medieval times, there was a French saying, *Il te faudra de vert vestir, c' est la livrée aux amoureux*, which means "You must dress in green, it is the livery [an old word for costume or uniform] of lovers."

EMERALD

Emerald is one of the most precious of all gems. It is the green variety of the mineral beryl, and its chemical description is beryllium aluminum silicate. The beautiful vivid green color of the stone is due to chromium and traces of iron. All true emeralds have faults called inclusions—if a stone looks too perfect it can be a sign that it is a fake. These inclusions mean that emeralds shatter easily; they are also susceptible to drying out, which is why most are treated with oil after cutting and shaping.

Archeological evidence suggests emeralds were mined in Ancient Egypt at least as early as 1500 BCE. In the Bible, specifically the Old Testament book of *Exodus*, emeralds are mentioned along with other precious stones as symbols of the tribes of Israel. Medieval beliefs about emeralds emphasized their protective powers. In the sixteenth century, when Spanish invaders reached Central and South America, they discovered the wealth of the Aztec and Inca empires lay in emeralds as well as gold. Emeralds were sacred to these peoples and were often carved into intricate jewelry or small goblets. Today, some of the finest emeralds come from Colombia; one of the largest ever found there is called the Gachala Emerald, which weighs 858 carats (6 ounces; 172 grams).

Gem-quality emeralds are expensive and are usually set in fine jewelry. Sometimes crystal suppliers sell opaque lower-grade emeralds, often called green beryl, which is much more affordable and can be used in crystal healing. Emerald makes a lovely gem remedy (*see pages 284–85*) that can be taken to help balance the heart and purify the emotions.

EMERALD IN BEDROCK

FACETED

EMERALD

FORM AND STRUCTURE
hexagonal system in varied
sizes and levels of clarity,
with inclusions

clearest specimens are gem
quality, more opaque ones are
lower grade

cutting and polishing reveals
the radiance of the finest stones

COLOR
deep vivid green

GEOGRAPHICAL SOURCES
Africa, Brazil, Colombia, Russia

RARITY
rare and expensive as gem
quality; more easily available
at lower grade

HARDNESS
7.5–8

PHYSICAL/EMOTIONAL USES
is directly associated with the
heart chakra, stimulating love
in its most pure expression

traditionally said to enhance
wisdom and inner knowing, and
to enhance love in a relationship

improves the eyesight and
supports liver function

HEALING EFFECTS
in healing layouts, place over the
heart, throat, or third-eye chakras
to facilitate clearer and purer
expression and feeling of love

PERSONAL USES
wear as jewelry or carry a piece
of opaque emerald with you
to bring love into your life

traditionally the birthstone for
the month of May

AVENTURINE

Green aventurine is a form of microcrystalline quartz containing particles of mica, a mineral that creates a sparkling sheen, so the stone glistens as it is turned to the light. All over the world its hardness has made it a popular choice for carving; in India—one of the main sources of green aventurine—it is made into beads, jewelry, boxes, and intricate figurines. Because it forms in massive blocks, this stone is also popular in interior and exterior mosaics, as flooring or internal cladding in combination with marble. In Russia, it was used after the 1880s by the Russian jeweler Fabergé, along with other semiprecious and precious stones—including jasper, lapis lazuli, quartz, jade, and sapphire—to make the famous Fabergé eggs.

Aventurine is an inexpensive and popular choice for starting a collection of crystals. Small polished stones have a smooth and silky feel. Darker specimens tend to have more mica particles, and lighter stones are more opaque. The name aventurine is from the Italian *a ventura*, which means "by chance"; this name is also given to goldstone, a form of sparkling Italian glass invented in the eighteenth century (*see pages 278–79*).

In crystal healing, green aventurine is used to clear away negative emotional patterns and support the heart chakra; it also calms powerful feelings in the lower abdomen, such as anger, centered on the solar plexus chakra. Aventurine encourages the expression of gratitude, hope, and positive approaches to life. It can also help to neutralize geopathic or electromagnetic stress.

POLISHED

AVENTURINE

FORM AND STRUCTURE
trigonal system, microcrystalline
quartz in massive formations

easily carved, and when polished
has a silky feel

COLOR
mostly green with specks of
sparkling fuchsite, also red, blue

GEOGRAPHICAL SOURCES
Brazil, India, Russia

RARITY
easily obtained

HARDNESS
7

PHYSICAL/EMOTIONAL USES
supports the system after a
period of illness—encourages
tissue repair and healing

supports the heart and
circulation

encourages optimism and a
renewed zest for life, as well
as gratitude for what has
been received

an excellent stone to destress
and calm body and mind

HEALING EFFECTS
in healing layouts, can be placed
over the heart chakra or the lower
abdomen to clear negativity and
encourage harmony

its energy is gentle and
supportive

PERSONAL USES
carry with you as a stone of
harmony

meditate with aventurine to take
stock of where you are and where
you would like to be

GREEN FLUORITE

Fluorite (calcium flouride) is a very common mineral that occurs in a whole range of different colors, and is popular as a first element in crystal collections. It often occurs in association with other minerals, such as quartz or calcite. Its cubic structure means that it forms very defined crystals with beautiful geometric shapes, often octahedral— a three-dimensional shape with eight facets. The name *fluorite* has been incorporated into a word describing one of its properties—this mineral will give off fluorescent light in different colors in the presence of ultraviolet rays. Industrially, the mineral fluorite is mined and used as a source of fluorine for fluorinated drinking water, as well as being part of the steel-and-glass manufacturing processes.

Though purple and rainbow—green-and-purple—specimens are most often associated with it, pure green fluorite is also easy to find as well as stunningly beautiful. It has a semiopaque appearance and a color similar to deep tropical seas—a deep, soft green with a hint of turquoise. It immediately calls to the heart chakra, its gentle soothing presence easing turbulent emotions. Meditating with it brings a sense of peace, like sinking into balmy healing water. It encourages deep relaxation and the freeing of the mind, which is always so busy controlling everything. Green fluorite reestablishes the heart as the feeling center through which life can be interpreted and sensed.

RAW

GREEN FLUORITE

FORM AND STRUCTURE
isometric system, forming cubic crystals sometimes in twinned pairs, typically also following an octahedral pattern

COLOR
green, also purple, white, yellow, pink, green-and-purple combined, black

GEOGRAPHICAL SOURCES
Brazil, Britain, China, Germany, Mexico, South Africa, Spain, USA

RARITY
easy to obtain

HARDNESS
4

PHYSICAL/EMOTIONAL USES
gently restores and reopens the heart chakra, especially if there has been any emotional stress or trauma

supports and encourages new growth and new beginnings

clears the mind and encourages creative thinking in line with the heart's deepest wishes

its balanced structure helps decision making where feelings need to be balanced with practical demands

HEALING EFFECTS
in healing layouts, place over the heart chakra to restore and enhance its energy; add rose quartz in the same location, bringing in the energy of unconditional love to heal emotional wounds

PERSONAL USES
place in a healing room to enable peace and a clear focus during meditation

place a piece in the bath to experience its relaxing effects

JADE

Jade is a name given to two very different minerals—nephrite, which is calcium magnesium silicate, and jadeite, which is sodium aluminum silicate. In the nineteenth century, when mineralogy developed more sophisticated identification techniques, these two types were able to be distinguished from each other; until that time they had both been called jade. Nephrite comes in creamy or pale green shades, whereas jadeite has more dazzling emerald green hues, along with variations in mauve, pink, black, or blue. Translucent emerald green jadeite is the most expensive and highly prized variety today. The name jade originally comes from Spanish *piedra de ijada*, meaning "stone of the loins," alluding to jade's reputed ability to heal the kidneys.

Jade has a long and fascinating history. In the Stone Age, it was used to make ax blades and weapons, being used as a slicing tool long before metal. As stone-cutting and polishing techniques improved—particularly in East Asian countries, such as China, and in Central America—jade was carved into statues, masks, ornamental breastplates, and jewelry. Fine examples of Mayan jade carvings can be seen today in museums in Guatemala. In China, priceless jade statues, often of dragons, decorated the palaces and tombs of the emperors. In New Zealand, the local jade is a stone of power in the traditions of the Maori.

CARVED JADE

RAW

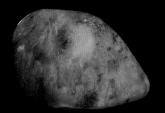

JADE

FORM AND STRUCTURE
nephrite and jadeite are examples
of monoclinic systems, and are
both tough stones with a tightly
bonded microcrystalline structure

COLOR
green in different shades, from
vivid emerald to pale green,
also mauve, pink, red, black,
and violet

GEOGRAPHICAL SOURCES
nephrite: Canada, India,
New Zealand, USA;
jadeite: Guatemala, Japan,
Myanmar, Russia

RARITY
nephrite is easier to obtain and
cheaper; top-quality emerald-
green jadeite is rarer and more
expensive

HARDNESS
nephrite 6; jadeite 6.5–7

PHYSICAL/EMOTIONAL USES
used to balance and support
the energy of the kidneys and
lymphatic system

calms the nervous system and
eases over-excitement

is linked emotionally with the
heart chakra, enabling the
sharing of unconditional love
with all beings

HEALING EFFECTS
place over the heart chakra to
open it and balance its energies,
or over the throat or third-eye
chakras to assist spiritual
expansion

PERSONAL USES
place in your home to encourage
peace and abundance

MALACHITE

Malachite is copper carbonate hydroxide, a distinctive, vivid green mineral with different shaded bands. It is popular with collectors and can be carved into many shapes, including spheres, dishes, and figurines. It occurs in amorphous-looking masses, sometimes as encrustations over or combined with other minerals, such as azurite or chrysocolla. It is always found near copper ore deposits. In the southwestern states of the USA, it is often set in silver jewelry as an alternative to turquoise.

The availability of massive forms of malachite has led to its large-scale use in impressive buildings, such as the Hermitage Palace, now a museum in St. Petersburg, Russia. There, the Malachite Room was given its name because of the stone features in the floors and pillars, and it includes a large, solid malachite vase.

Archeological evidence shows that copper ore and associated minerals, including malachite, were mined from around six thousand years ago at Timna, near Eilat in Israel. Excavations of the ancient workings there date back as far as the late Neolithic period. In powdered form, malachite and lapis lazuli were used as eye shadow by the Ancient Egyptians, and also as pigments in paint right up to the eighteenth century.

In crystal healing, malachite is valued as a powerful personal protector; it is a warrior stone, able to enhance passion for life and the energy to be true to one's unique purpose.

RAW

POLISHED

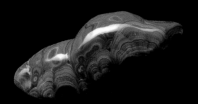

MALACHITE

FORM AND STRUCTURE
occurs as a monoclinic system in massive granular, botryoidal, or stalactitic forms, sometimes in combination with other minerals, such as azurite or chrysocolla

COLOR
vivid emerald green with distinctive bands

GEOGRAPHICAL SOURCES
Africa, Australia, Britain, France, Mexico, Russia, South Africa, USA

RARITY
easily obtained

HARDNESS
3.5–4

PHYSICAL/EMOTIONAL USES
is an important protective stone, particularly if you are affected by people who leech your energy

increases personal power and self-confidence, while enabling you to follow the true feelings of your heart

acts as a shield against electromagnetic disturbances

supports the blood and healthy circulation as well as improving resistance to disease

HEALING EFFECTS
place over the heart chakra to build and increase its energy, over the throat to enhance loving communication, or over the solar plexus for personal protection

PERSONAL USES
a large piece of malachite in the home helps neutralize environmental pollution

a small piece can be carried as personal protection, as well as to increase self-confidence

THE DARK GREEN RAY

This is the mysterious green of ancient trees. In the part of our minds the psychologist Carl Jung called the "collective unconscious," we share memories of a time when landmasses were covered in dense vegetation and humans lived on the edges of vast tracts of forest. The realm of the trees became associated with magic, mystery, and immortal beings who populated our stories and myths, including Merlin the magician, who was the teacher of the legendary King Arthur. Merlin, in the earliest version of the tales, is an elemental being, in touch with the powers of nature, able to merge with trees and disappear. In Norse mythology, there was a divine tree called Yggdrasil, which was a symbol of the whole world.

Dark green is a rich and yet opaque shade, like carpets of pine needles in glades of evergreen trees or thick moss covering boulders. In the deepest reaches of the forest, there is a kind of silence in the air, a feeling of deepest calm. The energy of the dark green ray takes us back to our roots and reminds us that those are in the earth. A wonderful tree that exemplifies this quality of darkest green is the Atlas cedar, *Cedrus atlantica*, towering majestically to over 120 feet (40 meters) in height, with a tent of dense branches sweeping right down to the ground. Standing beneath it gives a feeling of immense protection.

Restoring our connection to the earth

The slower vibration of dark green helps to calm and ground energies
that are overexcited and haphazard. Crystals and minerals with this color
are wonderful to use as anchors for the body's energy field. In the mad
rush of modern life, it is easy to lose track of yourself. Dark green stones
help to reestablish equilibrium, steadying the human energy field in
the midst of a host of different types of radiation and electromagnetic
interference. We sometimes forget that all the technology that surrounds
us is still new and we do not yet fully appreciate its long-term effects on
our physiology. For millennia, our bodies and energy fields resonated to
the earth in its natural state. Dark green energy helps to stabilize and
restore our connection to our planet.

We also need to rediscover and delight in mystery. In modern times,
we are pressurized to analyze everything to death and dismiss individual
experiences of reality. Dark green energy teaches that the realm of
mystery is still present if we care to explore.

MOLDAVITE

Moldavite is one of a group of amorphous glassy-looking crystals called tektites that are found in different parts of the world, including the USA, Australia, and Eastern Europe. Moldavite gets its name from the area of Moldauthein in the Czech Republic, close to an impact crater in the south of Germany where, some fifteen million years ago, a huge meteorite crashed into the earth. The formation of moldavite and other tektites is still under debate, but one theory suggests that the "splash field" which is generated by the impact of a meteor might have caused violent temperatures and dust clouds in the atmosphere, leading to the mixing and melting of airborne minerals, which then fell to the ground. This would account for the widespread scattering of deposits that tend to turn up as individual pieces or shards in the ground.

Moldavite is rare and even small pieces are surprisingly costly. It can be faceted and polished, but it tends to be set simply in silver or gold without shaping because its natural forms have such unusual and beautiful features. It is lightweight due to its glassy structure, and its distinctive dark green color is semitransparent when held to the light.

Moldavite was used to make arrowheads and small spearheads in prehistoric times; like obsidian (volcanic glass; *see pages 206–7*) it can be shaped into razor sharp points and these were used long before metal. Excavations in Austria in the early 1900s uncovered a famous statue, the Venus of Willendorf, and with it some pieces of moldavite, suggesting that as far back as twenty-five thousand years ago this green stone held a deep significance.

RAW

MOLDAVITE

FORM AND STRUCTURE
amorphous, glasslike structure,
a combination of silicon dioxide,
aluminum oxide, and various
other minerals

forms in shards, teardrop shapes,
or irregular spiked pieces which
are called hedgehogs

COLOR
moldavite is dark, leafy or mossy
green; other tektites are brown
or black

GEOGRAPHICAL SOURCES
moldavite: Czech Republic
(southern Bohemia); other
tektites: Africa, Australia,
Indonesia

RARITY
becoming rare as supplies
dwindle

HARDNESS
5.5–6

PHYSICAL/EMOTIONAL USES
moldavite is a crystal born of
immense transformative forces,
and is therefore very intense in
its effects

assists meditation or lucid
dreaming and enhances psychic
abilities

a powerful activator, so make
sure you are ready

HEALING EFFECTS
because of its potent energy,
in layouts this crystal is best
handled by a crystal therapist

PERSONAL USES
wear as a pendant or carry a
small piece with you—and watch
out for change

MOSS AGATE

Moss agate is a translucent form of chalcedony quartz, distinctive because it contains inclusions of dark green minerals that look just like moss or tiny lichens. They become particularly visible when the crystal is held to the light. These inclusions give the stone its name, although in a mineralogical sense it is not actually an agate because it is not banded like most true agates—carnelian, for example.

In India, which is an important source of the stone, moss agate is carved into beads and rounded cabochon shapes. Small tumblestones are pretty and inexpensive, making moss agate a popular choice for starting a crystal collection. Children are fascinated by this stone because it really does look as though there are plants inside it.

In crystal healing, moss agate helps to connect with the magical side of nature. In Indian traditional beliefs, tiny spirits live in the elemental world just beyond the visible spectrum and these are called devas. It is said that every single plant and flower has its own deva, attuned to its needs and able to communicate with it. Moss agate is a stone that helps to open perception to the spirits of nature that guard and protect the plant kingdom. It can be beneficial to ask for their help in your garden to keep your plants healthy; try placing a piece of moss agate outside as a gift to them. Wearing moss agate also helps you stay attuned to the benevolent abundance of nature, which always provides what is needed.

CARVED AND
POLISHED

POLISHED

MOSS AGATE

FORM AND STRUCTURE
trigonal system

chalcedony quartz with green or dark green inclusions of minerals, such as hornblende

silky smooth when polished

COLOR
pale translucent stone with deep green inclusions clearly visible

GEOGRAPHICAL SOURCES
India, Middle East, USA

RARITY
easily sourced, inexpensive

HARDNESS
6.5

PHYSICAL/EMOTIONAL USES
has a calming and stabilizing effect on the entire system and helps recovery after a period of illness

increases optimism and a sense of self-worth

eases emotional stress and strengthens the aura/energy field

HEALING EFFECTS
in healing layouts, place over the heart or solar plexus chakras to ease emotional stress and bring a sense of calm

use with rose quartz in the same locations in order to enhance unconditional love and spiritual expansion

PERSONAL USES
wear moss agate to promote abundance and self-confidence in your life, as well as awareness of the spiritual reality of nature

SERAPHINITE

Seraphinite is a specific variety of clinochlore, which has a complex chemical structure of magnesium iron aluminum silicate hydroxide. Clinochlore is one of a large group of minerals called the chlorites, which tend to form because of the metamorphic and hydrothermal alterations of iron and magnesium silicates in the earth's crust. The *chlore* part of the name is derived from the Greek word for green, since this is the main color of these minerals.

Seraphinite is a descriptive name given to a particularly beautiful type of clinochlore found solely in Siberia, Russia, in the area around Lake Baikal. It is so called because the deep green color of the mineral is interlaced with beautiful shimmering silvery strands that resemble wings—hence the name seraphinite, reminiscent of angels. The effect is enhanced by splitting the stone along its natural cleavage and polishing its vertical facets. This crystal is relatively new to Western Europe and is more expensive compared with other semiprecious stones. Seraphinite is soft and scratches easily, so it needs to be stored with care. It is becoming popular set in silver.

In crystal healing, seraphinite is used to heighten perception of the angelic realms. In these troubled times, the angels are needed, and many spiritual traditions revere them as messengers of peace and hope. Wearing or using seraphinite can bring your own guardian angel and spiritual guides close to you. It can also be used to align the "light body"—that is, the level of the human energy field that is closest to spirit—with the physical body, enabling an experience of humanity on all possible levels.

RAW

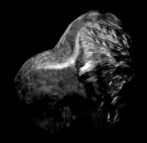

SERAPHINITE

FORM AND STRUCTURE
monoclinic system, forming in massive granular shapes, with typical white strands creating a shimmering luminescence

COLOR
deep green shot through with silvery white

GEOGRAPHICAL SOURCES
Russia

RARITY
easily obtained from specialist stores; expensive

HARDNESS
2–2.5 (store with care because it is soft)

PHYSICAL/EMOTIONAL USES
opens the crown chakra at the top of the skull, facilitating deeply spiritual experiences

is symbolic of angels and can be used to assist communication with your spirit guides

helps redefine your sense of self in line with the universe

HEALING EFFECTS
in healing layouts, place over the heart chakra or the crown chakra to assist spiritual expansion

use with lavender quartz to enhance angelic contact

PERSONAL USES
wear or meditate with seraphinite to experience the energy of your spirit at work in your body

THE BLUE-GREEN RAY

Blue-green is the color of the oceans that flood the earth's surface, turning it into the "blue planet." When astronauts stood on the moon and looked back at their home planet, one of the things that struck them most was the astonishing shade of the oceans, dwarfing the land masses and showing the earth as a planet of water. Life on earth began in the oceans; as organisms and creatures adapted to life on land, so over millions of years the earth was populated by its current inhabitants. The oceans are still vital to the survival of every living thing, and are part of the vast self-regulating ecological system that is our planet.

The color blue-green creates a feeling that draws us to the beach for vacations, a longing to swim in turquoise waters. We associate blue-green with freedom from stress: the simplicity of floating in water and relaxing releases us from anxiety. It is no accident that the most common color for tiles in a swimming pool is blue-green; even in artificially created water environments we seek to recreate the sea and a sense of carefree abandon. We watch television programs about the oceans, and the creatures that live in them amaze us with their grace, spontaneity, and beauty—dolphins leaping out of turquoise water, for example.

Healing qualities of blue-green

Energetically, blue-green is formed by the merging of the shade of the heart chakra (green) and throat chakra (blue). This particular energy frequency encourages love from the heart chakra to be communicated via the throat chakra. Coming back to the ocean, there are now many people who have swum with wild dolphins. This is often recounted as a profound experience of love, and it happens in the deep blue-green waters of the earth, the nourishing and nurturing environment of our planet.

In healing, blue-green soothes the upper area of the chest, easing physical problems, such as chesty coughs. There is a secondary energy center in the upper chest area called the thymus chakra, which helps to support the immune system. Its color is blue-green, symbolizing a bridge between the heart and the throat, allowing loving communication, which is one of the most powerful of all healing tools.

TURQUOISE

This beautiful mineral is so popular it also gives its name to the color turquoise, or hydrated copper aluminum sulfate, to give it its chemical name. Its finest grades are expensive, but cheaper lower grades are easily obtained. Turquoise occurs in different shades of blue-green, according to its origin: Iranian or Persian turquoise has a deep blue-green color; Middle Eastern deposits tend to be more green; turquoise from the southern states of the USA varies from duck-egg blue to deeper shades, often mottled with inclusions of pyrite or other minerals.

Turquoise has been mined since at least 6000 BCE, with the oldest known mines being located in the Middle East. In Muslim countries, such as Persia (modern-day Iran), turquoise was used in many different ways, from elaborate jewelry to inlays in boxes, furniture, mosaic floors, and walls. The popular use of turquoise spread from the Middle East to India, where it features in the famous Taj Mahal. In Central America, it was a very popular stone with the Aztecs, who used it with gold, malachite, jet, jade, and other minerals to create sacred objects such as masks and ceremonial knives. The ancient Native American tribe called the Anasazi, of Chaco Canyon, prospered thanks to their trading of turquoise artifacts, and to this day the Pueblo, Navajo, and Apache tribes use turquoise in silver jewelry and sacred amulets. In China, turquoise has been carved into jewelry and statues for over three thousand years. The Ancient Egyptians also used it extensively in necklaces, rings, beads, and the carving of scarabs (the sacred beetles associated with the god Khepri).

RAW

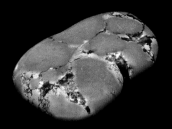

TURQUOISE

FORM AND STRUCTURE
triclinic system, occurring in massive forms, microcrystalline in nature so it looks opaque

waxy in luster with a silken feel to touch

COLOR
blue-green in various shades from light to dark

GEOGRAPHICAL SOURCES
Afghanistan, Australia, Middle East, USA

RARITY
easily obtained, although high-grade turquoise is more expensive

HARDNESS
5–6

PHYSICAL/EMOTIONAL USES
a stone linked to powerful personal protection, for which it has been used for millennia

enables communication from the highest source of love

eases feelings of negativity and low self-esteem

supports and energizes the thymus chakra over the breastbone, boosting immunity

HEALING EFFECTS
in healing layouts, place over the heart, throat, or thymus area to ease breathing and open the flow of emotional expression

PERSONAL USES
wear over the upper chest as personal and psychic protection

LABRADORITE

Labradorite belongs to a group of minerals called feldspars that are commonly found in the earth's crust. It has a dull, dark blue-green appearance, but when it is turned to the light, particularly if the surface is polished, a deep blue sheen appears, a shimmering light effect called labradorescence. Most labradorite has this blue sheen; some examples show green, violet, or orange effects, and rare specimens a rainbow of all these colors—these are called spectrolite. The color display occurs because of intergrowths of minerals laid down in parallel patterns within the crystal structure. These bounce light back and forth, creating the shimmering effect. Labradorite's chemical description is calcium sodium aluminum silicate.

The name labradorite refers to the region of Labrador in Canada, where the stone was first found in the eighteenth century. However, it is now obtained from many other countries, including Finland, Norway, Mexico, and the USA. In the nineteenth century, discoveries of deposits in Russia increased its popularity in jewelry.

In healing, labradorite is considered a powerful stone of magic and transformation, awakening supersensory abilities and the awareness of other states of energy. It is a stone with relevance to shamanism, the most ancient form of spirituality on the planet, where healing is sought in parallel planes of existence by traveling between different states of consciousness. Labradorite's silken sheen symbolizes the mysterious elusive qualities of these other realms of existence.

RAW

POLISHED

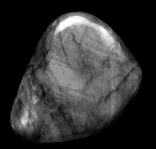

LABRADORITE

FORM AND STRUCTURE
triclinic system, found in blocks or chunks embedded within low-silica igneous rocks, sometimes in twinning formations

COLOR
dull blue-green, with a mostly iridescent blue sheen when turned to the light, also orange, green, violet; rainbow iridescence is found in specimens called spectrolite

GEOGRAPHICAL SOURCES
Canada, Finland, Mexico, Norway, Russia, USA

HARDNESS
6–6.5

PHYSICAL/EMOTIONAL USES
increases extrasensory perception and may stimulate recall of past lives, clairvoyance, or other psychic abilities

activates a sense of multiple levels of consciousness

protects the wearer from any negative influences

HEALING EFFECTS
in healing layouts, place over the top of the head or hold a piece in each hand to facilitate expanded awareness

balance with a piece of smoky quartz between the feet in order to stay anchored

make a gem remedy (*see pages 284–85*) with it to support new spiritual growth

PERSONAL USES
wear to sense your expansion into new levels of yourself

BLUE-GREEN APATITE

Apatite is the name given to a group of minerals that have a calcium phosphate base, but with different levels of fluorine, chlorine, or hydroxide in their matrix. This causes a variety of colors to form, from deep blue-green to green to yellow, or even pink or violet. Blue-green apatite is attractive because it is granular when polished, showing a mixture of microcrystalline fragments as well as more opaque layers. Some samples even have specks of rutile, a mineral causing a sparkling effect. Apatite is another common mineral, which forms in igneous, sedimentary, or metamorphic rocks; it sometimes forms large terminated points with a clearer appearance, but is more often found in microcrystalline masses. The name comes from the Greek "to deceive" because it is easily mistaken for other minerals—the pale green form, for instance, looks similar to peridot. However, apatite is soft and this helps to distinguish it from harder, more precious gems.

Apatite is also found in human tooth enamel, which shows that human bodies contain the same minerals that are found in the structure of the planet on which we live. The foods we eat, from plants through to animals that feed on them, provide the building blocks for our physical framework—we are literally made of the same stuff. Ancient cultures, such as the Native American peoples, believed that rocks and crystals were, on a certain level, alive. They sensed the link between the mineral kingdom and human existence, understanding and respecting the qualities we share.

RAW

APATITE

FORM AND STRUCTURE
hexagonal system, in granular, massive, microcrystalline formations or, more rarely, in terminated points

mottled appearance

COLOR
blue-green, green, yellow, pink, violet, clear

GEOGRAPHICAL SOURCES
Canada, Germany, Mexico, Russia

RARITY
obtain from specialist suppliers

HARDNESS
5

PHYSICAL/EMOTIONAL USES
helps mend broken bones or ease problems with the teeth; eases recovery after physical injury/trauma

structures thinking and brings disordered ideas into a more cohesive focus

helps concentration and the solving of problems

expands awareness of other levels of consciousness, such as the mineral kingdom itself

HEALING EFFECTS
place a stone in each hand to balance both sides of the body

place it over the heart chakra, or over the breastbone (thymus chakra) to strengthen the energetic structure of the aura

place over the third-eye chakra between the eyebrows to expand spiritual perception

PERSONAL USES
carry with you to help you focus when you have to multitask

AMAZONITE

Amazonite is potassium aluminum silicate, a member of the large group of minerals called feldspars. These make up a large proportion of the earth's crust and are found all over the world. Amazonite gets its name because of its distinctive blue-green color, which is supposed to echo the rich green hues of the Amazon River and surrounding rainforest. However, deposits of this green feldspar are unknown in the Amazon region, and the best specimens come from Colorado, where it often occurs in clusters with smoky quartz, and also the Ilmen Mountains in Russia, where it is found alongside granite deposits. Recent analysis of amazonite has revealed that the blue-green color is caused by the reaction of lead and water with its feldspar structure. It forms in large chunky pieces, mottled with lighter green specks, and it has a silky feel when polished.

Amazonite has been used to make jewelry for centuries. In Ancient Egypt, as well as being used in collars and rings, it was also carved into bigger tablets to be engraved with sacred writing. Today, the relics of Pharaoh Tutankhamun reside in the Egyptian Museum in Cairo. The museum provides a description of the famous gold mask of the young king as containing amazonite inlays in the collar, along with lapis lazuli and carnelian. Amazonite was also popular in ancient Indian jewelry, especially when set in gold. Its moderate hardness makes it suitable for carving, and even today it is cut and polished into rounded cabochon shapes to highlight the color, before being set in rings, pendants, or earrings.

RAW

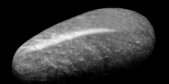

AMAZONITE

FORM AND STRUCTURE
triclinic system, forming in large blocks or chunks; a green variety of feldspar

COLOR
blue-green, mottled with paler green specks

GEOGRAPHICAL SOURCES
Madagascar, Russia, USA

RARITY
easily obtained

HARDNESS
6

PHYSICAL/EMOTIONAL USES
facilitates loving, compassionate communication by bridging the heart and throat chakras

clears congestion from the chest and throat areas, easing sore throats and coughs

helps to balance the energy of the thyroid gland in the throat

helps recovery from emotional disputes or traumas

HEALING EFFECTS
place over the heart, upper chest, or throat areas as needed to soothe and balance the energy of these chakras; use your sensitivity to gauge which area needs most help, or if you are working with a friend, ask them to sense as well

PERSONAL USES
wear or carry to help you deal with emotional challenges and relationship issues

CHRYSOCOLLA

Chrysocolla is an interesting mineral that occurs in a wide variety of shades, ranging from green to blue-green to turquoise. Its chemical description is hydrated copper silicate. It is often found in deposits alongside other green stones, such as malachite or blue azurite, and like malachite, it can form amorphous shapes, sometimes botryoidal (grapelike) or stalactites. It is also found in oxidized layers surrounding copper ore. Chrysocolla is extremely soft and, therefore, not used in jewelry making unless it is embedded in harder microcrystalline quartz, which allows some polishing. It is then known as gem silica.

In crystal healing, chrysocolla links the energy of the heart chakra upward toward the throat, the center of expression. It helps to facilitate clear emotional expression and the communication of the heart's desire. A powerful physical healing stone, chrysocolla supports the function of the thyroid gland as well as easing throat problems. The greenish blue color of chrysocolla reminds us of pictures of the earth from space; meditation with this stone helps build a strong connection with our planet as a spiritual being, sometimes referred to as Gaia. This Ancient Greek name for the earth refers to a Mother Goddess, one who supplies all the support that her children need. In modern science, the Gaia hypothesis explains how the earth is a completely interactive ecological system of which we are an integral part, and the beautiful shade of green found in chrysocolla reminds us that our planet is a living thing.

POLISHED

CHRYSOCOLLA

FORM AND STRUCTURE
orthorhombic; occurs in massive
amorphous forms, encrusted over
other rocks or in stalactites or
botryoidal clusters, often in
conjunction with other stones,
such as malachite

COLOR
green, blue-green, turquoise

GEOGRAPHICAL SOURCES
Russia, USA, Zaire

RARITY
easily obtained

HARDNESS
2.5–3.5

PHYSICAL/EMOTIONAL USES
clearing, cleansing, and
detoxifying, chrysocolla supports
the immune system and helps
rebuild energy in the body after a
period of illness, especially chest
infections

supports the metabolism and
enhances the function of the
thyroid gland

on an emotional level chrysocolla
assists in the expression of
personal truth and helps those
who speak in public

HEALING EFFECTS
in healing layouts, place over the
breastbone or the throat to clear
toxicity and rebuild energy after
a period of illness

place over the heart chakra to
ease breathing

place under the base of the skull
to open psychic perception

PERSONAL USES
place in your workspace to
encourage clear communication

meditate with a stone to increase
your creativity and intuition

THE PALE BLUE-GREEN RAY

Previously, when we discussed the pale green ray, we discovered that lighter versions of colors are higher and more subtle vibratory levels of that particular energy. Here we meet pale turquoise, most beautifully illustrated by the stone aquamarine. This is the color of tropical seas, whose pale blue-green waters brushing white sands make us think of paradise.

Legends from antiquity tell the tale of Atlantis, a fabled city-kingdom on several interconnected islands, reputed by some to have been sited in the Atlantic Ocean thousands of years ago. Its inhabitants grew into a highly evolved society, skilled in all manner of arts. They were reputed to have been experts in working with metals and crystals; their buildings, according to Ancient Greek accounts of their distant history, were decorated with gold, silver, and orichalcum (most likely a type of bronze). They were also fabled astronomers, scientists, and healers. The story of Atlantis is a fascinating one, and continues to be a mystery to this day. The destruction of Atlantis occurred as a cataclysmic eruption, either a massive volcanic explosion or perhaps, as more esoteric commentators suggest, because of overambitious experiments by Atlantean scientists.

A mysterious realm

For a long time, Atlantis was a place of mysteries and marvels, and many crystal healers feel an affinity with its energies, which are illustrated by this pale turquoise color. The "atl" root in its name is also found in the Inca and Aztec languages of the Americas, and it means water. Bathed by the seas, its islands interlaced with water, Atlantis is a symbol of a mystical kingdom just beyond reality. Legend also has it that one day Atlantis will rise once more.

Some people have a belief that this life is not the only one we live, and many religions around the world teach that life is a lesson we come back to experience many times over. It is possible to sense and even remember past lives if your mind can open to this. Many healers connect with Atlantis as a source of previous therapeutic experience, either by using energy or by using crystals. It is said that powerful crystal healing was practiced in Atlantis, and this was transferred to Ancient Egypt after the destruction of the island kingdom. Using or wearing pale turquoise stones may help you connect to this lineage if that is your heritage. Meditating with the pale turquoise ray is a deeply gentle and subtle portal to this mysterious, wonderful realm.

AQUAMARINE

Aquamarine is a pale blue-green variety of the mineral type called beryl, a group that also includes emeralds. This is a hard gemstone, colored by trace amounts of iron that have been absorbed into the crystal structure. It forms in long hexagonal crystals, often embedded in pegmatite, with parallel striations, looking like lines running up its vertical structure. Its name shows its link to the water kingdoms—*aqua* is the Latin for water, and marine means "of the sea." True aquamarine has a beautiful translucent pale turquoise color. It is popular cut and faceted in jewelry; many commercially available stones are made from lower-grade beryl, heat-treated to produce a pale blue color. Today Brazil is a leading source of natural high-quality aquamarine, as is Pakistan, where wonderful specimens are also found. For the mineral or crystal collector, slightly more greenish colored specimens of uncut aquamarine are available and are relatively inexpensive.

In crystal healing, aquamarine is a vital tool in cleansing and clearing the water element, a symbol of our emotions and feelings. Our bodies are over 70 percent water in their makeup and, therefore, keeping this element balanced is crucial. We breathe in the air element, physically we are made of minerals from the earth element, and we transmit sparks of energy—or fire—between every one of the millions of nerve endings in our bodies. Yet the water element is the most relevant, perhaps because originally we evolved from creatures who lived in the sea. Aquamarine helps to cleanse the water element within, clearing away emotional toxicity and bringing inner peace.

RAW

PALE VARIATION

AQUAMARINE

FORM AND STRUCTURE
hexagonal system, forming large bladed crystals with vertical striations and the tendency to break along that vertical plane; often found embedded in igneous rocks

COLOR
pale turquoise blue, some specimens greener; faceted stones often heat-treated low-grade beryl

GEOGRAPHICAL SOURCES
Brazil, Pakistan

RARITY
easily sourced as small pieces or tumblestones, or as faceted gems in jewelry

HARDNESS
7.5–8

PHYSICAL/EMOTIONAL USES
an excellent crystal to boost the thymus chakra, enhancing immunity and keeping the bridge clear between the heart chakra and throat chakra

encourages creative verbal expression and harmonious communication

soothes powerful feelings, such as grief or loneliness

cools infection and reduces inflammation

HEALING EFFECTS
place raw stones over the breastbone or at the throat to facilitate these chakras and enhance communication

hold over any areas that are inflamed or painful

PERSONAL USES
wear to enhance communication and enable your highest truth to be expressed in the world

CHRYSOPRASE

Chrysoprase is a type of chalcedony quartz with a microcrystalline structure. Its color comes from nickel deposits, and it can be found in pale shades through to brighter green. Raw pieces have a granular appearance, similar to pieces of coconut, and a waxy sheen when polished. Chrysoprase is one of the rarer forms of quartz, so finer specimens are more expensive. It is cut into cabochons for jewelry making to bring out the color. If struck in a particular direction, it fractures into flakes in the same way as flint, so care has to be taken when working it. However, chrysoprase can be finely carved, and good pieces can even be mistaken for jade. The color fades in strong sunlight, so store it with care.

For hundreds of years, the only major source of chrysoprase was a region of southwestern Poland between the Czech Republic and Germany, formerly known as Lower Silesia. This area has a complex geology and is extremely rich in all kinds of mineral deposits, including gold, silver, serpentine, quartz, marble, granite, alabaster—and chrysoprase. Stone artifacts from as far back as the Iron Age show the long history of Silesian carving and masonry. In the eighteenth century, Frederick II, King of Prussia (now part of Germany), conquered Lower Silesia. In particular, he wanted its deposits of green chrysoprase, which he used to decorate the halls of his favorite palace at Potsdam near Berlin. Today, fine examples of chrysoprase are also found in Australia, Brazil, and Madagascar.

POLISHED

CHRYSOPRASE

FORM AND STRUCTURE
chalcedony quartz,
microcrystalline structure
in a trigonal system

raw pieces look chunky and
granular and the stone is waxy
when polished

COLOR
pale greenish turquoise to vivid
apple green

GEOGRAPHICAL SOURCES
formerly Poland; currently
Australia, Brazil, Madagascar

RARITY
easily obtained as tumblestones

HARDNESS
7

PHYSICAL/EMOTIONAL USES
supports and expands the heart
chakra, easing negative feelings
and promoting joy

it attracts new love and
abundance and helps to prepare
the system for new phases in life

brings restful sleep and aids
general relaxation

HEALING EFFECTS
place over the heart chakra,
throat, or third eye as required to
soothe the system and enhance
the flow of positive feelings

a piece in each hand helps to
balance left and right sides of
the body

PERSONAL USES
carry or wear to attract new
relationships into your life

put a piece under your pillow to
improve the quality of your sleep

THE PALE BLUE RAY

Think of a summer sky, a pale limpid blue, stretching to the horizon. This shade of blue is soft, expansive, and cool, soothing to the eye and calming to the mind. In a world filled with stress, noise, bustle, pressure, and too many people, this pale blue ray is an effective antidote. Not surprisingly, it features on many vacation brochures because a wide blue sky promises freedom, space, room to breathe, calm, and relaxation.

One of the most wonderful things about flying is the moment when the aircraft breaks through the cloud layer and suddenly your eyes are dazzled by the vastness of the sky and its clear blue color thousands of feet above the ground. You realize that there is so much space up there, far more than your mind normally considers while it is buried in the nuts and bolts of everyday life. The pale blue ray can be used in meditation to expand and clarify your thought processes; it encourages a wider bird's-eye view of problems instead of getting buried in minute details. The Ancient Egyptians revered the falcon, a bird that seems to live in that wide blue sky; they made that bird a god and called it Horus, the one who sees everything from above. In the Andes, the Peruvians assigned similar qualities to the condor, one of the largest winged creatures on earth, revered as a bringer of wisdom and clear vision. To take a wider view of any problem, ask yourself: How would this look from above? If I could see it from the air, what would I perceive? The pale blue ray encourages a change of perspective, which can often highlight aspects of a situation that have previously been ignored.

A calming influence

In color healing, pale blue is a higher frequency of deep blue, a more subtle vibration, no less powerful but with a different intensity. It is calmer and more soothing, gentle in its effects on body and mind. It dissolves inflamed emotions and anger and eases inflammation on the body. It is linked to the throat chakra, the place of communication, and as a subtle vibration of blue it encourages higher levels of awareness. These are faculties such as clairvoyance—which literally means clear seeing—or other kinds of extrasensory perception, which it is possible to develop by meditating with the pale blue ray or using pale blue crystals.

BLUE LACE AGATE

Agates are a type of quartz found in many different variations, colors, and effects. They are microcrystalline in form, meaning that their structure is too small even to be seen through a microscope. They are usually opaque in texture and form in large masses, rather than in large defined crystal points as is the case with pure quartz. Blue lace agate is a descriptive name, referring to the delicate bands, stripes, and specks of cream interlaced with pale blue within this semiprecious stone. Polished blue lace agate tumblestones are easy to find in crystal stores and are inexpensive; they are very popular early choices for crystal collectors, particularly children who enjoy having a variety of colored stones. Larger pieces can be particularly striking, with wavelike creamy bands contrasting with the soft blue color, especially if polished.

Blue lace agate's color echoes the sky, bringing a sense of peace and calm. It is an excellent stone to place in a meditation or healing space to enhance the atmosphere. It can also be used as a focus for meditation to expand the mind into new realms or to bring relaxation. In the bath, it can be used with rose quartz to create a balance of love and peace within the water; this helps dissolve daily worries and cares. Taking a gem remedy made with blue lace agate (*see pages 284–85*) brings a beautiful sense of peace and calm to the whole system, especially if there is any emotional trauma or severe anxiety.

RAW

POLISHED

BLUE LACE AGATE

FORM AND STRUCTURE
microcrystalline quartz in a trigonal system, with bands confirming it as an agate

usually found in large massed formations

COLOR
pale sky blue, with bands of cream or white

GEOGRAPHICAL SOURCES
Brazil, South Africa, Uruguay

RARITY
easily obtained as small tumblestones or larger pieces

HARDNESS
7

PHYSICAL/EMOTIONAL USES
brings peace and calm to the mind and helps the body to relax

enhances sleep quality and eases feelings of anxiety

helps to ease anger and inflamed emotions

eases inflammation on the body, as well as throat problems, such as soreness or coughs

HEALING EFFECTS
place over the throat chakra to enhance communication and soothe the effects of anger

hold a piece in each hand to relax completely the whole body

place over the third-eye chakra to soothe the mind and enhance clear thinking

PERSONAL USES
wear close to the throat to facilitate clear and gentle communication

BLUE MOONSTONE

Moonstone is a type of feldspar made up of potassium aluminum silicate. It occurs in different types, but all have a special kind of sheen called labradorescence, also found in labradorite (*see pages 124–25*). This is a light effect caused by parallel hairlike structures laid down within the crystalline structure that bounce light back and forth at a certain angle, so creating different colored effects. Some types of moonstone have a pearly white sheen and this one—called blue, or sometimes rainbow, moonstone—produces a beautiful pale blue shimmer when the stone is turned to the light. It has a long history of use in jewelry—it is particularly effective when used with silver, an increasingly popular setting today—and the Celts and Romans wore it set in rings, brooches, or pendants.

Moonstone is a descriptive name for this feldspar, given because the stone itself recalls the shimmering changing presence of the moon in the sky. In crystal healing, moonstone is regarded as a symbolic stone for enhancing feminine energy. This does not mean it is only for women. All human beings experience male and female energies—other ways of describing them are dynamic and receptive, active or passive. We all need to have an outward "doing" (masculine) focus sometimes, and an inner "being" (feminine) one at others. Whenever there is a need to bring peace and calm within, then blue moonstone is a good option to choose.

POLISHED

BLUE MOONSTONE

FORM AND STRUCTURE
feldspar with a monoclinic structure, forming in blocks or tabular structures; often found in conjunction with granite

COLOR
pale or clear base with lamellar intergrowths which create a luminous blue sheen; also white or gold sheen

GEOGRAPHICAL SOURCES
Australia, India, Myanmar, Sri Lanka, USA

RARITY
easily obtained

HARDNESS
7

PHYSICAL/EMOTIONAL USES
gently calms the energy field or aura, clearing negative energies and strengthening inner peace

eases feelings of stress and anxiety, especially those linked to hormonal fluctuations such as in PMS or menopause

enhances intuition by increasing receptivity to your inner creativity

HEALING EFFECTS
place three stones over the heart, throat, and third eye to experience deep peace and inner calm, and open the mind to deeper reflection

hold a piece in each hand to enhance receptive feminine energy in the body

PERSONAL USES
wear over the heart or throat chakra to enhance inner peace and tranquility

CHALCEDONY

Chalcedony is another type of microcrystalline quartz with a softly opaque appearance caused by fibrous patterns in its structure; this is more evident when the stone is polished. Unlike jasper, which is more granular, chalcedony has a smooth almost silky feel as well as a gently luminous hue. Blue chalcedony has a very particular shade, a soft yet intense pale blue similar to bluebells in a spring wood.

Blue and other shades of chalcedony have been used since antiquity in carving. In Ancient Egypt, this was a favorite stone for the making of scarabs—models of the dung beetle associated with the god Khepri, who was a god of birth and renewal. These were carried as protective amulets. The Ancient Greeks also used the scarab shape—an oval like half a hard-boiled egg with one rounded and one flat surface. Artists—notably Epimenes, who lived around 500 BCE—carved delicate crouching figures into the flat surface with clearly visible features, including musculature, faces, and even hair, although the pieces were no more than 2 inches (5 centimeters) in diameter. In Roman times, chalcedony was carved into beautiful rings, necklaces, and cameos with detailed features.

Blue chalcedony is used in crystal healing to help the eyes, soothing soreness and clarifying vision. It enhances inner peace and restores balance in body and mind. It also dissolves any negativity around the throat chakra area, restoring the ability to communicate clearly.

POLISHED

CHALCEDONY

FORM AND STRUCTURE
microcrystalline quartz in a
trigonal system, occurring in
stalactitic, botryoidal (grapelike)
or large masses

COLOR
vivid pale blue, also purple

GEOGRAPHICAL SOURCES
Brazil, Czech Republic, India,
Madagascar, Mexico, Turkey

RARITY
easily obtained

HARDNESS
7

PHYSICAL/EMOTIONAL USES
opens the throat chakra to
enable the truest forms of
self-expression

also eases throat problems,
including soreness or coughs

acts as a soothing balm to cool
the deepest emotional wounds

eases fears and nightmares,
bringing peace and serenity

soothes the lungs and assists
with breathing problems,
especially if related to stress

HEALING EFFECTS
in healing layouts, place on
the throat chakra to clear any
communication blockages, or
to ease soreness or loss of voice

place over the third-eye chakra to
open and widen inner perception

PERSONAL USES
place under your pillow to help
troubled sleep

wear over your throat or carry
it with you in order to ease
communication issues

KYANITE

Kyanite is an unusual mineral composed of aluminum silicate. It forms in parallel flat blades with visible striations, which are often embedded in metamorphic rocks or veins of quartz. An unusual characteristic of kyanite is that it has two levels of hardness—it is relatively weak when scratched parallel to the vertical axis and harder if scratched across that axis. It is usually found in a pale blue color—sometimes in deeper blue or white specimens—and it also has an attractive pearly sheen on its flat surfaces. Viewed from the side, a piece of kyanite resembles sheets of thin paper laid on top of each other, rather like a book. It has a tendency to splinter along the vertical axis, which means it is not often set in jewelry, and needs to be handled and stored with care.

The distinctive structure of kyanite makes it a useful tool in crystal healing as a conductor of energy, working like a channel to transmute negativity out of the system and allow positive energy in. It opens up new avenues of awareness and enables deeper levels to be accessed in meditation. It allows energy to be in a constant flow, moving through any blockages with ease; it is a classic crystal for letting go, releasing old patterns that no longer serve us.

BLADE

RAW

KYANITE

FORM AND STRUCTURE
triclinic system, forming long, flat parallel blades with striations

mostly found embedded in metamorphic rocks, such as quartz, gneiss, or schist

COLOR
mostly pale blue, sometimes vivid blue, also white, gray, black

GEOGRAPHICAL SOURCES
Africa, Brazil, Eastern Europe, India, Russia, Switzerland, USA

RARITY
obtain from specialist suppliers

HARDNESS
4.5 when scratched parallel to the vertical axis; 6.5 if scratched perpendicular to it

PHYSICAL/EMOTIONAL USES
works like a conductor, helping energy move through the body, taking out negativity and bringing in positive vibrations

eases away old mental and emotional blocks, enabling a new perspective

activates and enhances psychic abilities

helps heal broken bones by directing healing energy through the trauma site

HEALING EFFECTS
place over a trauma site to ease the flow of healing energy

place over the solar plexus chakra to bring upper and lower chakras into alignment as a true energy conduit

place over the third eye to enhance psychic awareness

PERSONAL USES
wear over the chest to keep energy flowing smoothly throughout the system

CELESTITE AND ANGELITE

Celestite is a beautiful pale blue crystal composed of strontium sulfate; it is also known as celestine. Celestite forms in granite and pegmatite rocks, in geodes, where clusters of crystals form inside a rocky outer layer; it also forms beautiful clear terminated points. The best quality sky blue celestite comes from Madagascar; examples from the USA are more whitish blue and form large solid-looking points. Ground powder from celestite produces a bright red flame when it is burned; this is due to the strontium in its makeup. Although celestite is a translucent mineral, it is heavier than quartz. It sometimes occurs with deposits of bright yellow sulfur, making a sky blue-and-yellow contrasting combination that is much sought after by mineral collectors.

Another sulfate mineral with a very similar sky blue color—although it should not be confused with celestite—is commonly known as angelite. This is calcium sulfate, officially called blue anhydrite. It forms because of water loss in gypsum, which shrinks over time, making the structure of anhydrite more opaque and compact. Blue anhydrite/angelite does not have the translucent quality of celestite, but it is equally beautiful and can be found as polished tumblestones.

Both celestite and blue anhydrite/angelite are used in crystal healing to bring a gentle uplifting healing energy to a space, connecting to subtle levels of energy and one's angelic guides.

RAW ANGELITE

POLISHED ANGELITE

CELESTITE

FORM AND STRUCTURE
celestite and angelite are formed
in an orthorhombic system

celestite can produce large
translucent crystal terminations
as well as geodes or clusters;
angelite forms in masses

COLOR
both pale sky blue; most celestite
is translucent, apart from
samples from the USA with a
whiter hue; angelite is opaque

both stones are also white, gray,
and violet

GEOGRAPHICAL SOURCES
celestite: Europe, Madagascar,
USA; angelite: Mexico, Peru

RARITY
both are easily obtained

HARDNESS
3–3.5 for both

PHYSICAL/EMOTIONAL USES
both minerals are used to open
the throat chakra, enabling
higher levels of communication
(both transmitting and receiving)
with one's spiritual guides and
angelic helpers

the soft sky blue color of these
stones is gently supportive and
encourages awareness of being

HEALING EFFECTS
both work with the third-eye and
crown chakras to open spiritual
perception, and can be placed
there in healing layouts

PERSONAL USES
place geodes of celestite in a
room to bring in gentle healing
energy; carry angelite with you
to remind you of your guides

149

THE SAPPHIRE BLUE RAY

This is the deep sapphire blue seen in many medieval paintings and stained-glass windows, the translucent shade most famously featured in the cathedral of Chartres in France. The interior of this vast stone building is kept deliberately dim, so that when light passes through the windows, the depth and beauty of the blue glass is particularly noticeable. Medieval glassworkers used ground lapis lazuli to stain glass that particular color, and artists used the same powder in mixing paint to show the Virgin Mary's robes.

The world-famous illuminated manuscript, *Les Très Riches Heures du Duc de Berry*, produced in France in the early fifteenth century, is full of wonderful paintings showing this lapis-lazuli blue in the skies, the backgrounds, and the rich clothing of the nobility. From medieval times through to the Renaissance, this shade of blue was a symbol of heaven. This ray is also traditionally associated with Archangel Michael, whose mantle is often depicted in that color. Michael is a warrior-archangel with a flaming sword, a symbol of the sparkling awakening of the soul. Meditating on the deep blue ray is a wonderful way to connect with the angel's energy.

The color of self-expression

Crystals and minerals with this deep blue color have been worn since antiquity by kings, potentates, and spiritual leaders. Today they can be worn and possessed by everyone; they help to awaken leadership qualities within you, showing you the nature of your unique gifts. It is easy to "hide your light under a bushel," but every person has a unique quality, and wearing deep blue crystals helps you gain confidence in communicating whatever your individual gifts may be. Sapphire blue is the luminous shade of the throat chakra, the place of self-expression. This means tapping into your creativity, bringing into the world the gifts you hold, enabling them to serve a higher purpose. Imagine a world where creativity and individual expression were encouraged and truly valued in every child and adult. How might that world be? We admire the geniuses in our history books, but the ability to create lies in each and every one of us. The sapphire blue ray connects us to our deepest dreams and helps us make them come alive.

LAPIS LAZULI

Lapis lazuli is a complex mineral made up of several key compounds, the main one being lazurite (25 to 40 percent), which is composed of sodium, aluminum, silicon, oxygen, sulfur, and chlorine. In addition, lapis lazuli typically contains pyrite (which look like specks of gold), and calcite (whitish specks), as well as sparkling inclusions of mica. It is classed as a rock because of this complex structure. The name comes from *lapis*, meaning stone in Latin, and *lazuli*, Medieval Latin derived from the Ancient Persian *lazhward*, meaning blue. The deep sapphire blue color of this stone is due to the sulfur in its matrix. It was first mined in Afghanistan several thousand years ago, and although other geographical sources exist, the Afghan mountains still yield some of the finest specimens.

This stone has been prized since deepest antiquity, with the Ancient Egyptians being the most famous people to work it into jewelry, sacred regalia for the pharaohs, rings, scarab carvings, boxes, mosaics, and other ornaments. The color of lapis lazuli is even more dramatic when it is polished, and it is relatively hard, allowing it to be carved. Egyptian artifacts made of lapis lazuli show the skill of these ancient jewelry experts in bringing out the color. In powdered form, lapis lazuli was also used by Egyptian nobility as an eye shadow, along with black kohl to outline the eyes.

The medieval paint pigment called ultramarine, used in glassmaking and painting, was originally made with ground lapis lazuli, until the invention of artificial pigments after the Renaissance.

RAW

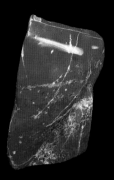

LAPIS LAZULI

FORM AND STRUCTURE
isometric rock formation, mostly
as an aggregate of many minerals
in combination, mainly lazurite

COLOR
vivid sapphire blue with
inclusions of golden pyrite
and whitish calcite

GEOGRAPHICAL SOURCES
Afghanistan, India, Myanmar,
Pakistan, Russia, USA

RARITY
obtain from specialist crystal
suppliers

HARDNESS
5.5

PHYSICAL/EMOTIONAL USES
lapis lazuli is a royal stone,
activating the higher chakras
of the third eye and crown of the
head, bringing awareness of soul
and spiritual purpose in life

its golden inclusions of pyrite
connect it to solar energy, so it
also balances the solar plexus
chakra (individual will) with
the crown chakra (divine will)

promotes courage and lets
your light shine in the world

soothes migraines and
headaches

HEALING EFFECTS
place over the throat, third eye
(between the eyebrows), or
crown—whichever feels most
appropriate—to facilitate
connection with soul purpose

PERSONAL USES
wear over the chest or throat
to bring you self-confidence

153

SODALITE

Sodalite is made up of sodium aluminum silicate chloride; it gets its name from its sodium content. It is a rich, royal blue color interspersed with white veins; good-quality pieces when polished look similar to lapis lazuli. However, it does not contain the pyrite inclusions found in the latter. Another way to tell the two apart is that sodalite is more brittle; clear lines in its structure show potential breaking points. It rarely forms crystals, and it is usually found in massive granular formations or fillings in veins in igneous rocks. One source of rare but particularly fine crystalline sodalite is the region around the volcano Vesuvius in southern Italy; massive deposits are also found in Ontario, Canada. Today, sodalite is used with jasper, marble, and other fine masonry stones to create elaborate interiors, staircases, and floors for important buildings, its blue color providing an attractive contrast.

Sodalite is used in crystal healing to expand the mind and encourage clear thinking; it is also a good stone to work with if you are trying to establish a regular meditation practice. It helps to connect your physical self to your spiritual self so that you can move easily between different levels of consciousness. Its deep blue color brings peace and tranquility to your environment and soothes unruly emotions. It also enhances your intuition, that sixth sense that encourages creativity and spontaneity.

POLISHED

RAW

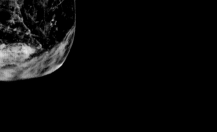

SODALITE

FORM AND STRUCTURE
cubic system in massive or granular formations, rarely crystalline

often occurs in igneous rocks

when polished has a waxy feel

COLOR
mostly royal blue, also white, gray, green

GEOGRAPHICAL SOURCES
Canada, Greenland, Italy, Myanmar, Russia, USA

RARITY
easily obtained as small tumblestones

HARDNESS
6

PHYSICAL/EMOTIONAL USES
calms the mind and helps relaxation

assists the third-eye chakra to focus in meditation

helps support the immune system, clearing problems in the throat and sinuses, and soothes the voice, especially if the vocal chords are strained

helps clear feelings of internal conflict or lack of certainty, encouraging self-confidence in personal expression

HEALING EFFECTS
place over the throat or third eye between the eyebrows, whichever feels appropriate, to clear negative feelings and increase relaxation in the system

place over the heart chakra with a piece of rose quartz to create a fusion of intuition and unconditional love

PERSONAL USES
meditate with it to improve your ability to let go of everyday concerns and focus on just being

BLUE SAPPHIRE

Sapphire is a form of aluminum oxide also known as corundum, one of the hardest of known minerals apart from diamond. Blue sapphire is colored by titanium and iron; it is the most common color associated with this stone, although sapphires can be found in many other shades, including pink. Some of the finest sapphires come from Sri Lanka, India, and Myanmar. Like diamonds and rubies (the red form of corundrum), gem-grade sapphires are also expensive. Most blue sapphires are heated to remove any brown or gray pigments, and this improves the color saturation. Some blue sapphires show a phenomenon called asterism, where intersecting needles of rutile cause the appearance of a six-rayed star pattern in the stone; this is visible if it is shaped and polished in a cabochon or rounded dome. These stones are rare and even more costly. However, low-grade pieces of sapphire are available to use in crystal healing, often in small hexagonal-shaped slices, and these are inexpensive to buy.

The word *sapphire* is from the Hebrew *sapir*. In the Old Testament, in *Exodus*, Chapter 28, Verse 18, where God is giving instructions for making a breastplate for Aaron the high priest, sapphire is one of twelve sacred stones required to complete the design. The breastplate has four rows of three precious stones; the second row contains emerald, sapphire, and diamond. These stones are all symbols of the twelve tribes of Israel.

FACETED

BLUE SAPPHIRE

FORM AND STRUCTURE
trigonal system, forming
prismatic, rhombohedral,
or tabular crystals, typically
six-sided

mostly found in igneous or
metamorphic rocks

COLOR
most commonly royal blue,
also pink, clear, yellow, green,
multicolored

GEOGRAPHICAL SOURCES
Australia, Afghanistan, Africa,
India, Myanmar, Pakistan,
Sri Lanka

RARITY
top-grade gem sapphires are
rare and expensive; lower-grade
pieces can be obtained from
crystal suppliers

HARDNESS
9

PHYSICAL/EMOTIONAL USES
enhances self-confidence and
standing in the world, due to its
long history as a badge of royalty

links the throat, third-eye, and
crown chakras, opening spiritual
and psychic perception

enhances clear and truthful
communication

soothes inflammation and burns

eases headaches and migraines

HEALING EFFECTS
place pieces over the throat,
third-eye, and crown chakras to
help expand psychic awareness

use a gem remedy made with
blue sapphire (*see pages
284–85*) in order to relax the
system and ease tension

PERSONAL USES
wear to enhance self-confidence

157

AZURITE

Azurite is copper carbonate hydroxide, a mineral with an intense blue color—in fact, its name derives from *azure*, meaning sky blue. It is often found embedded in quartz, calcite, or in combination with malachite.

The distinctive bright shade of blue of azurite is caused by specific chemical reactions between copper, carbonates, hydrogen, and oxygen. The chemistry of azurite is almost the same as malachite, and these two minerals are often found together, which can indicate the presence of other rich copper sulfide ores nearby. Azurite, if exposed to weathering on the earth's surface, can be replaced by malachite in a process called pseudomorphism—the replacing of one form by another. Freshly mined azurite starts to oxidize into malachite due to the presence of moisture in the air; it is therefore important to keep azurite specimens in a dry environment so that moisture doesn't cause the color to change.

In powdered form azurite has been used as a paint pigment for centuries. It turns green when mixed with oil, or gray when combined with egg yolk; a deeper blue shade was traditionally obtained by heating the mineral gently—although overheating turns it black. However, because of azurite's tendency to oxidize, when medieval painters used it to fill in a blue sky, over time it would turn green!

It is possible to buy pieces of combined azurite and malachite; this helps to balance the energies of the heart (malachite) and throat (azurite), enabling clear and loving communication.

RAW

AZURITE

FORM AND STRUCTURE
massive, nodular forms in a monoclinic system, sometimes stalactitic, borytroidal, or encrusting forms

can be found embedded in other rocks such as quartz, or in combination with malachite (appearing as contrasting deep blue areas)

COLOR
deep sapphire blue with a tendency to grow lighter or more green if exposed to ultraviolet rays or excess moisture

GEOGRAPHICAL SOURCES
Australia, Chile, China, Mexico, Namibia, Russia

RARITY
rare and expensive; obtain from specialist crystal suppliers

HARDNESS
3.5–4 (soft, so handle with care)

PHYSICAL/EMOTIONAL USES
its stunning deep blue color immediately links it to the throat and third-eye chakras, enabling communication of inner wisdom and enhanced perception of one's inner guides

enhances intuition and psychic abilities

stimulates creativity

helps clear throat problems and supports the voice

HEALING EFFECTS
place on the throat, over the top of the head, or hold a piece in each hand to balance and enhance spiritual awareness

PERSONAL USES
place in your environment to help inspire you and increase your creativity

THE DARK BLUE RAY

Dark blue is the deep shade of the night sky, the background to the map of the stars that has fascinated human beings throughout time. When the moon is full and the sky is clear, the stars gleam against a vast dark blue backdrop that stretches away into infinity. This nocturnal blue shade deepens the further away you get from cities; with no background interference from street lighting, the contrast with the stars is even more beautiful. In the middle of the wilderness, such as in the Australian outback or on the plains of Africa, the night sky sheds its own light from millions of stars. In some traditions, the night sky is a map leading to the realm of the ancestors.

This deep dark blue that draws the eye toward the unknown realms of the cosmos is a symbol of mystery and the infinite; it is the color of the third-eye chakra that sits between the eyebrows. In Sanskrit, the name of this chakra is *ajna*, which means "to perceive,"—this chakra is the doorway to expanded perception both beyond and within oneself. It is an energy hub that opens during deep meditation to enable the dissolution of the "lower self"—that driven by the desires and needs of the physical body—and the opening of the "higher self," which is in tune with spiritual wisdom. It takes many years of dedicated meditation practice to reach this state, but the way is open to all. Sometimes the activity of this energy center is experienced as flashes of intuition or déjà vu. Sometimes this is also called the "sixth sense." It points to levels of perception beyond the rational mind.

Open your mind

Using deep blue crystals on the body or in meditation can modulate your energy field and help you experience a different state of being, one that is free from stress and restriction. In this state, you learn that your consciousness is a much wider entity than simply your mind, which wants to analyze and control your experiences. Your consciousness is a wider sense of yourself, one that raises your awareness of your place in the greater scheme of things. If you have never come across this idea before, try taking some meditation classes, where a teacher can help you explore what it may mean to you.

IOLITE

Iolite is a popular and unusual deep blue gemstone with a translucency that has earned it the descriptive name of water sapphire. However, unlike true sapphires, which are a type of corundrum (aluminum oxide), iolite is magnesium aluminum silicate. It displays pleochroism—that is, different shades of color according to the light—which can be violet, yellow, light blue, or gray reflections within the stone. It is relatively hard and can be cut or faceted for making jewelry—it is particularly popular set in silver. Iolite is a gem variety of a mineral called cordierite, which occurs in metamorphic rocks. Iolite's magnesium content is an identifying factor, as other types of cordierite are rich in iron. Cordierite minerals are often found in granite deposits.

Iolite's translucent deep blue color is used in healing to facilitate inner journeying. This is a technique from shamanic tradition that enables a person to travel within using their imagination as a tool to help shed light on issues or problems. Iolite helps to open the third-eye chakra, which is the gateway to inner knowledge and understanding. It is soothing and peaceful, helping to dissolve any fear or anxiety and allowing the flow of information that is available to come to you. Any realizations that come from such inner journeying can have profound effects on your life. It is best to undertake such a journey in the company of a professional healer who knows how to guide you through it and help you stay centered while the process unfolds.

RAW

IOLITE

FORM AND STRUCTURE
orthorhombic system, forming
hexagonal prismatic twin
crystals, embedded granular
structures, or masses within
metamorphic rocks

COLOR
deep blue or violet-blue, with
pleochroism—color reflections—
in yellow, light blue, gray

GEOGRAPHICAL SOURCES
Canada, Madagascar, Myanmar,
Sri Lanka, Tanzania

RARITY
rare; obtain from specialist
crystal suppliers

HARDNESS
7–7.5

PHYSICAL/EMOTIONAL USES
activates the third-eye chakra,
enabling inner visions and
enhancing the clarity of dreams

soothes headaches and
migraines and eases eyestrain

calms overwrought nerves and
eases emotional stress

helps cure insomnia caused by
mental stress

HEALING EFFECTS
in healing layouts, it has a
calming and soothing effect on
the body

place over the third eye to clear
and open this energy center, or
over the throat or the heart to
balance the emotions and the
nervous system

PERSONAL USES
carry or wear iolite jewelry in
order to help calm your system
and ease stress

163

BLUE TIGER'S EYE

Tiger's eye is a form of silicon dioxide (quartz) with a special luminous quality in its structure. This occurs when fibers of a mineral called crocidolite are laid down and oxidized in parallel bands within the quartz structure. This creates a silky-looking shimmering impression when the stone is turned to the light, an effect that resembles a cat's eye, which is why it is called chatoyancy after the French word *chat*, meaning cat. Blue tiger's eye has many tightly packed crocidolite strands in its structure, giving it an opaque texture and a deep blue sheen when polished.

Tiger's eye has been a popular semiprecious stone for thousands of years, from the early civilizations of Mesopotamia and the early dynasties of Egypt. Different shades have been set in precious metals and used in the making of necklaces, rings, and torques, or inlaid into elaborate collars. Since Roman times, tiger's eye has also been a traditional choice for carving signet rings with individual crests or emblems to be used as stamps for sealing wax.

Although yellow tiger's eye (*see pages 72–73*), with its bands of gold to yellow-brown, is the most common form of the stone, the blue variety is just as beautiful with its shimmering blue layers. It is a stone of mystery, linked to the third eye and psychic perception. It has a more calming effect than the yellow variety, helping to ease mental anxiety and slow down an overactive brain.

POLISHED

BLUE TIGER'S EYE

FORM AND STRUCTURE
trigonal crystal system, forming
in large masses, with a typical
banding in the structure formed
of crocidolite fibers

COLOR
soft dark blue

GEOGRAPHICAL SOURCES
Australia, India, South Africa,
USA

RARITY
easily obtained as polished
tumblestones

HARDNESS
7

PHYSICAL/EMOTIONAL USES
soothes frontal headaches
caused by mental stress and
calms the system, easing anxiety

has a protective effect, shielding
the aura from unwanted
influences

opens the third-eye chakra and
facilitates lucid dreaming, as well
as conscious clear thinking

HEALING EFFECTS
in crystal-healing layouts, place
over the third eye to open that
chakra, or place a piece in each
hand to calm and balance the
whole system

works in harmony with rose
quartz placed over the heart
chakra to soothe the emotions
and encourage self-expression
with love

PERSONAL USES
wear or carry it to ease stress
and anxiety, and facilitate clear
thinking

take a bath with it to recover
from anxiety

THE DARK PURPLE RAY

Dark purple is a rich color that has been esteemed since ancient times as a mark of high rank and even royalty. Archeological evidence from Crete, home of the Minoan culture, has shown that extraction of what came to be known as Tyrian purple was taking place there as far back as 2000 BCE. Tyrian purple was a dye from the mucous membranes of particular mollusks, obtained using a process that even modern methods have been unable to replicate. In Roman times, the city of Tyre in the eastern Mediterranean was a key

producing area, hence the name Tyrian.

Apart from yellow saffron, this purple dye was one of the most expensive and rare coloring ingredients of the ancient world. In consequence, it was reserved for the robes, cloaks, and togas of persons of highest rank, and even though it disappeared from use over time, superseded by plant dyes such as woad or mallow, the color purple remained a sign of high status. To this day purple still has a regal quality about it.

This rich purple shade is also associated with the crown chakra, the energy center that sits at the top of the skull. This is often represented in religious paintings as a golden light or halo around the top of the head. Purple is the color that activates it. The meaning of the crown chakra is that each and every one of us has the opportunity to achieve our highest potential on all levels—mental, physical, emotional, and spiritual—during our lifetime.

Experience an awakening

Activation of the crown chakra is a crowning moment in life, signaling the awakening of divine purpose in an individual. Unlike in the past, when such potential was reserved only for a few, now there is recognition that such awakening can be experienced by all. Moments of ecstatic understanding, flashes of blinding intuition or inspiration, bursts of enlightened creativity that seem to come from nowhere are all signs of the crown chakra opening. These moments need to be balanced by being firmly grounded. A wise Buddhist saying states: "Before enlightenment, chopping wood and carrying water; after enlightenment, chopping wood and carrying water." As human beings, we operate on all levels, physical as well as spiritual. The key thing is balance.

Purple crocus flowers in the spring illustrate this deep color so well; they contrast with the rich shade of new grass, showing the abundance of heart-chakra green against the rich purple of crown-chakra wisdom.

AMETHYST

One of the best known and most beautiful of all crystals, amethyst occurs in a great variety of shades, shapes, and sizes. It is a form of quartz, colored by different amounts of iron or aluminum. It occurs in deep dark purple shades right through to pale lilac, depending on where it is found. It can be found in large defined crystal points or in clusters of smaller terminations. Another typical and large form of amethyst occurs inside pockets (vugs) within volcanic rock, where an enclosed space has been created as the rock cooled. These formations have a greenish encrustation on the outside, and when split open reveal hundreds of amethyst crystals growing toward the center of the space. Entire pieces are available—and expensive—to buy; they look like crystal caves, and can be as much as 3 feet (1 meter) in size.

The word *amethyst* is derived from an Ancient Greek term meaning "not drunk"; the Modern Greek word *amethystos* still means both amethyst and sober. Quite why the stone was given this name is unclear, although its deep purple color resembles wine. Amethyst has been prized since ancient times, used in royal jewelry such as crowns and scepters, and in the Old Testament book of *Exodus* it is mentioned as one of the stones in the breastplate of Aaron the High Priest.

POLISHED

AMETHYST

FORM AND STRUCTURE
trigonal crystal system, forming long prismatic crystals with a six-sided pyramid at the point; also smaller clusters or large, enclosed masses (vugs) within metamorphic rocks

COLOR
dark vivid purple, in a variety of shades, to pale lilac

GEOGRAPHICAL SOURCES
Africa, Brazil, Canada, Mexico, Russia, USA

RARITY
easily obtained in a variety of shapes and sizes

HARDNESS
7

PHYSICAL/EMOTIONAL USES
popular as a purifying crystal, helping to ward off negative influences and protect from environmental stress

calms the nervous system and the brain, easing headaches and migraines

opens the third-eye and crown chakras, enhancing personal spiritual awareness

HEALING EFFECTS
in crystal-healing layouts, place over the top of the head or on the third eye to expand conscious awareness

place a circle of eight pieces of amethyst around the body to create a protective field

PERSONAL USES
wear or carry amethyst to help calm mental stress

place under the pillow to improve the quality of sleep

SUGILITE

Sugilite is a rare mineral with a complex structure—potassium sodium lithium iron manganese aluminum silicate. A Japanese geologist called Ken-ichi Sugi discovered the first specimens in 1944, and the mineral was named after him. Sugilite has a distinctive rich purple color, and forms in masses rather than in defined crystals. It is very opaque but lustrous when polished; this highlights the deep purple color that indicates the best specimens. Originally found in Japan, sugilite has also been discovered in Canada as well as South Africa, which is the main source today. As a recently discovered mineral, it is highly sought after by collectors.

In crystal healing, sugilite is viewed as a stone for modern times, especially since it was only discovered in the middle of the twentieth century. Its deep violet color is seen as a sign of awakening spiritual potential through the crown chakra, and an influx of wisdom to humans dedicated to caring for the planet in the midst of all the changes we are currently experiencing. It brings divine love to the earth and encourages compassion for all beings. Instead of the jumbled nightly downloads of everyday life we call dreams, sugilite encourages lucid dreaming, a process of reframing and creating the future. Spiritual shamanic teachings from Peru and the Aborigines of Australia suggest that the future reaches out to touch us in such dreams, and all we have to do is perceive the signs and act upon them.

POLISHED

SUGILITE

FORM AND STRUCTURE
hexagonal crystal system, mostly forming in massive deposits, rarely as tiny prismatic crystals

COLOR
mostly rich purple; also brown, yellow, pink, black

GEOGRAPHICAL SOURCES
Japan, Canada, South Africa

RARITY
quite rare; obtain from specialist crystal suppliers

HARDNESS
6–6.5

PHYSICAL/EMOTIONAL USES
opens the third-eye and crown chakras, enabling deep perceptions of one's place on the earth and spiritual destiny

clarifies and intensifies dreams

helps the human body act like a conductor, bringing spiritual energy to the earth and returning physical energy to the cosmos

eases insomnia, especially that caused by mental overload

brings positive feelings of optimism and hope for the future

HEALING EFFECTS
in healing layouts, place it over the crown of the head to facilitate the opening of that chakra; place a piece of smoky quartz at the feet to help bring the energy down the body

PERSONAL USES
wear or carry in order to help you maintain your focus on your own spiritual destiny

THE LIGHT PURPLE RAY

Light purple is a more subtle vibration of purple.
In shade, it resembles the color of lavender flowers;
on lavender farms in the region of Provence, France,
the whole landscape is covered with that gentle purple
shade, soothing to the eye and the senses. The sight
and the scent of the flowers immediately calms and
relaxes body and mind. From a healing perspective,
light purple can be used to calm anxiety and fraught
emotions; it helps the mind let go of mental stress and
repetitive thought patterns.

The color of light purple or violet is also associated with the light of the
sun; ultraviolet—or UV—is an area of the color spectrum that humans
cannot see. However, every single day our bodies, eyes, and minds are
bathed in those rays. Somehow we know this, and we crave the light, so
when the weather is dark and miserable we feel the same. When the sun
comes out, our spirits lift and we want to be outside, to feel the sun on
our faces. This makes us feel better, stronger, more positive. Of course,
too much exposure to high levels of ultraviolet rays can be damaging,
but in moderation sunlight helps maintain health. It has positive effects
on our pituitary and pineal glands, and therefore the balance of
hormones in the system.

The violet flame

Violet is a gentle shade and is deeply soothing to the entire system.
Used in color healing to destress and calm overwrought nerves, it also
neutralizes anger and inflammatory emotions. In a spiritual context,
it is sometimes referred to as the violet flame, which is linked to the
alchemical concept of transforming base elements into precious metals.
In this case the flame is not orange-red as in physical fire, but rather
violet, a symbol of the transformation of the lower nature of humankind
to higher spiritual values. The violet flame is a tool that can be used in
meditation to dissolve away the burdens of the past and free the spirit
to open to the present moment—the point of choice and transformation.
To achieve this, try visualizing a fire with violet flames during meditation,
and then see yourself stepping into it. As you are completely bathed by
the flames, feel yourself being freed and renewed.

Light purple or violet stones—pale amethyst, for example—can be
used to help this process of transformation, bringing in the energy of the
violet ray, bathing the body, and preparing it to make new beginnings.

CHAROITE

Charoite is an unusual mineral with an extremely complex chemical structure. It looks like a form of marble, with streaks of light purple alternating with gray, white, and black specks, and sometimes contains inclusions of a gold-colored mineral called tennesite. Its chemical description is hydrated potassium sodium calcium barium strontium silicate hydroxide fluoride. It was discovered in the 1940s and named for the Chara River in the Sakha Republic, Siberia, Russia. The swirling patterns of lavender-purple mixed with the other colors in charoite are only visible when the stone is polished. It forms when limestone layers are infused with alkaline-rich minerals in the presence of heat. Charoite has a unique look and luster; it is rare because Russia is the only known source. It is relatively soft and can be made into jewelry, but it scratches easily and needs to be handled carefully.

Charoite's swirling, flowing appearance indicates it as a tool for transformation, working to cleanse and release old patterns and bring in new energy. The colors in the mineral are significant. The soft purple shade, the richest color, signifies spiritual opening; the black specks the transmutation of negative influences. Silvery streaks link to cleansing emotions and a lunar influence, and specks of gold signify input of solar radiance. Charoite has it all, in a sense, and among crystal healers it is seen as a powerful amplifier of personal energy and power, cleansing, grounding and preparing the way for spiritual expansion.

POLISHED

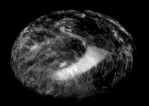

CHAROITE

FORM AND STRUCTURE
monoclinic structure of fibrous interlocking crystal masses formed by intrusions of alkaline-rich minerals into limestone layers under the influence of heat; found in large deposits

COLOR
lavender-purple mixed with black, gray, white, or gold-colored inclusions

GEOGRAPHICAL SOURCES
Russia

RARITY
fairly rare; obtain from specialist crystal suppliers

HARDNESS
5–6

PHYSICAL/EMOTIONAL USES
amplifies the energy of the crown chakra and has a powerful energetic cleansing effect in the system, clearing old memories, past traumas, and elements that no longer serve

helps the assimilation of higher-energy frequencies into the physical body

activates an awareness of "soul purpose," clarifying direction in life

HEALING EFFECTS
place charoite over the crown chakra to activate and amplify its energy

add rose quartz to the heart chakra to open to unconditional love, and a piece of smoky quartz at the feet to ground the flow of energy

PERSONAL USES
place in your personal space to increase your understanding of your "soul purpose"

PURPLE FLUORITE

Fluorite is a mineral found in many different colors, but particularly in a beautiful purple; fine specimens even rival amethyst in shade and luminosity. Fluorite is calcium fluoride and forms naturally in clusters of cubic crystals. In its natural state, it has a glasslike appearance and is semitransparent; it is even more lustrous when polished, and this brings out bands and variations in the color. It is thermoluminescent, glowing when heated, and also displays fluorescence, meaning it glows under ultraviolet light. It has four identical planes of cleavage, so when a rough piece is struck at the correct angles it will form a perfect diamond-shaped octahedron. Many lower-grade pieces sold in crystal shops have been shaped this way. Fluorite is a source of fluoride, used in toothpaste and in drinking water. It is a very common mineral and a popular choice when starting a crystal collection because there are so many variations of shape, color saturation, and shade to be found.

Fluorite's balanced geometric structure is a clue to its use in crystal healing—it is a stone that symbolizes changing chaos into order. It is helpful to have around if your life is in need of structure, if you have difficulty prioritizing, or if you feel scattered in all directions. Its straight edges and defined planes encourage logic and clear patterns of thinking.

POLISHED

PURPLE FLUORITE

FORM AND STRUCTURE
cubic structure producing cube-shaped crystals, transparent to translucent, sometimes twinned

often occurs with calcite, quartz, pyrite, or other sulfide minerals

COLOR
soft purple, also reddish orange, green, blue, pink, yellow; sometimes mixtures of colors in one stone (rainbow fluorite)

GEOGRAPHICAL SOURCES
Brazil, Britain, Canada, China, Germany, Mexico, Spain, USA

RARITY
easily obtained

HARDNESS
4

PHYSICAL/EMOTIONAL USES
encourages structured thinking and helps to solve mental difficulties

eases confused thought patterns and balances the mind

opens the third-eye chakra and enhances mental focus

helps center the mind during meditation practice

helps to heal the bony structure of the body, so is useful after any physical trauma

HEALING EFFECTS
place over the third eye to help calm and sort out disordered thinking, or place a piece over an injured area to help energize and rebalance it

PERSONAL USES
place in your workspace to help you concentrate

hold a piece during meditation to help center your mind

THE LILAC RAY

Lilac is a very delicate and subtle shade, a mixture of pale purple and pink, like the color of the sky at dusk. It is a tender and gentle color ray with an energy that soothes and nurtures the spirits. It is also cooling and soft, an antidote to the harsh stresses of the outside world. Being enveloped in lilac energy is like being in a cocoon, a safe place of inner peace.

The color combination that makes lilac is highly significant. The light purple vibration, as we have seen, is cleansing and clarifying to the mind. The pink ray is the ray of unconditional love, signifying a deep compassion for all living things. Lilac brings those two energies together and creates a bridge, allowing a peaceful mind to coexist with a loving heart. This creates a different lens through which to view the world, and a new potential experience of other people.

Lilac is also a beautiful energy ray to use as a nurturing tool to support the inner child. This is a level of consciousness that exists within everyone, whatever his or her physical age may be. In its most positive aspect, it is a space of innocence, spontaneity, joyfulness, and bubbling creativity; however, in many adults that place has been damaged, often through emotional troubles that reach back into childhood. Negative reinforcement and criticism from an early age may well have damaged the inner-child energy and left it depleted, to surface as angry or jealous emotional reactions when the adult personality feels threatened. A lot of psychotherapeutic work relates to rediscovering and healing the inner-child level, and the lilac ray can really help to reestablish a sense of that space as a safe and nurturing inner environment.

A supportive energy

Lilac-colored crystals include pale amethyst and the lighter shades of fluorite, as well as those featured in this section. These crystals can be laid in circles around the body to create a stable and safe energy field, especially if a person feels vulnerable and unable to talk. Sometimes inner feelings are difficult to express, and in that situation the application of a healing tool such as a crystal layout can give the body and mind the supportive energy that is needed in that moment. This can pave the way for further development in the healing process over time. Sufficient rest, relaxation, and breathing within a safe-energy environment are always extremely beneficial.

179

SPIRIT QUARTZ

This is an unusual member of the quartz family, made, like all quartz, of silicon dioxide, but having a unique structure. It forms well-defined columns with a faceted tip, and on the sides of each column there are hundreds of miniature quartz crystals giving off an attractive sparkling shine. Clusters of spirit quartz grow in a huge variety of shapes, sometimes resembling fingers. The crystals are mostly pale amethysts, although rare examples of white or yellow (citrine) spirit quartz have been found. Spirit quartz is a newcomer to the crystal-collecting world, discovered as recently as 2002, and is only found in South Africa, in an area approximately 50 miles (90 kilometers) northeast of Pretoria.

In crystal healing, the structure of spirit quartz causes great interest because the multidirectional and multifaceted nature of the mineral suggests it can activate many degrees of human potential. In personal terms, this means being alive, awake, and aware on many different levels. It works on the physical, emotional, mental, and spiritual planes at the same time, balancing and restructuring the energy field. This is a stone of transformation and a symbol of being all you truly are. Spirit quartz is also a symbol of human evolution and potential; just as the earth continually releases more and more unusual minerals, so human beings have infinite possibilities of choice and evolution. Spirit quartz symbolizes the diversity as well as the unifying aspects of humankind, whether at a family, community, or even broader level.

CLUSTER

SPIRIT QUARTZ

FORM AND STRUCTURE
trigonal system, producing
central columnar shapes with
defined termination points and
sides covered with tiny faceted
quartz crystals

COLOR
lilac purple, also, more rarely,
white or yellow

GEOGRAPHICAL SOURCES
South Africa

RARITY
fairly rare; available from
specialist crystal suppliers

HARDNESS
7

PHYSICAL/EMOTIONAL USES
multifaceted spirit quartz is like
a community within one stone;
it helps to heal group issues
or family concerns, easing
communication and conflict

on a personal level it activates all
the chakras and brings them into
balance with the outside world

helps create a feeling of stability
and a peaceful center in life

HEALING EFFECTS
in healing layouts, a piece of
spirit quartz placed either over
the top of the head or between
the feet will help align and
balance the entire energy field

PERSONAL USES
place in your home to encourage
peaceful and harmonious family
or group relationships

181

LAVENDER QUARTZ

This is an unusual and special form of rose quartz, named for its very specific lilac-pink shade. To appreciate fully the color variation, place a piece of ordinary rose quartz beside it, and then the difference between the two will become apparent. Lavender quartz has a beautiful gentle energy and hue. It is found mostly in Brazil, sometimes in large clusters or points. It responds well to carving because, like all quartz crystals, it is hard. Stunning carved pieces in the shape of a flame are becoming popular. Like rose quartz, lavender quartz can be semitransparent, clear, or opaque. Because it is less easy to find, it is more expensive than rose quartz, which is more common.

The color of lavender quartz is a combination of pink and light purple. Pink is the color of the higher level of the heart chakra, encouraging trust and opening to inner feelings as well as compassion. Light purple encourages the gentle opening of the crown chakra, bringing inspiration and wisdom. Lavender quartz is a gentle stone to work with in the practice of meditation; it encourages slow and patient spiritual unfolding rather than sudden transformational leaps. It helps to nurture the emotions, assisting in establishing a safe and peaceful space within. It symbolizes peace and tranquility. Placed under the pillow it helps to ease disturbed sleep and encourage deeper rest. It is a good stone to use with children to help them calm down after periods of physical activity.

POLISHED

LAVENDER QUARTZ

FORM AND STRUCTURE
silicon dioxide in a trigonal structure, forming in massive deposits or as clusters

COLOR
lilac-pink

GEOGRAPHICAL SOURCES
Brazil

RARITY
rare; obtain from specialist crystal suppliers

HARDNESS
7

PHYSICAL/EMOTIONAL USES
soothes the emotions and eases inner vulnerability

helps create a sense of peace and stability during times of stress

helps insomnia, nightmares, and irregular sleep patterns

eases headaches and migraines caused by mental overload

nurtures the inner-child energy and encourages trust and spontaneity

HEALING EFFECTS
in healing layouts, place over the heart to soothe the emotions, or the third-eye or crown chakras to ease mental stress or headaches

place a piece in each hand to balance the energies of the right and left sides of the body

PERSONAL USES
use with rose quartz in the bath to ease stress and tension

prepare a gem remedy with it (*see pages 284–85*) to help emotional stress

THE PINK RAY

The pink ray is the healing color of love, gentle and enveloping. It is another color associated with the heart chakra—we have already seen the effects of the green ray on this. Pink is a more subtle frequency of heart-chakra energy, yet very expansive; it symbolizes connection, opening, and fruition.

The simplest way to connect the green and pink rays is to visualize a rose, opening as a pink flower on a bed of green leaves. The color, shape, texture, and depth of the flower are attractive to the eye, and the rich, soft aroma lulls the sense of smell. It is almost impossible to walk past a stunning rose without stopping to admire it, taking a moment to appreciate it in all its beauty. The symbol of the rose has a long history in Western esoteric tradition—for example, it is linked with Aphrodite, the Greek goddess of love, and, later, with the Virgin Mary, the mother of Christ. Giving a rose as a symbol of love is a simple yet profound gesture, linking the perfection of the flower with the deep green of the leaves, symbols of growth.

Pink is also the color associated in healing with unconditional love. This is love uncluttered by expectations or desires, just simple feeling, open and inclusive. Letting go of expectations and allowing life to flow is one of the biggest lessons in healing. Unconditional love is not weakness or sentimentality; it is true compassion. It can dissolve old hurts and change the habits of a lifetime. The pink energy of the heart chakra helps facilitate personal experience of this kind of love. It is not limited to relationships, although it can enhance them; it goes beyond the demands of physical life, establishing a place of inner connection with the spirit.

The ray of compassion

One way to work with this energy, especially if you have relationship problems or persistent disagreements with someone, is to visualize them in a pool of pink water. Being bathed in the pink ray of unconditional love really does wonders for a situation, and it is worth trying this technique. Sometimes in the middle of a situation it is enough just to think "pink pool." As soon as you apply compassion to a problem, no matter how complex it seems, it is amazing how the energy changes. Another idea would be to give someone a pink stone, such as a piece of rose quartz, as a symbol of the pink ray of unconditional love.

ROSE QUARTZ

Rose quartz is one of the most popular of all crystals and often one of the first to be collected. It is a member of the quartz family that is colored by impurities of manganese or titanium. The range of color in rose quartz varies from deep to very pale pink. It is mostly fairly opaque, although some clearer polished pieces have a beautiful translucent appearance. Some rose quartz crystals also contain specks of rutile, a mineral that creates sparkling specks in the stone. Although as a quartz this mineral is classed as hard, its crystalline structure is often flawed, creating cracks or visible fractures, and this makes it challenging to cut. However, carved and polished pieces are available as well as natural rough stones. Rose quartz forms in massive deposits or in clusters. Small pieces are inexpensive, and are available as polished tumblestones.

In crystal healing, rose quartz is a popular tool, whether as uncut stones or as polished wands for directing energy. Its soothing frequency allows it to interact with the body and the energy field gently, bringing a nurturing and supportive element into a healing layout. Although its color links it directly to the heart chakra, there is no need to stick to this rigidly—rose quartz can be placed on the body wherever the energy of love is needed. Holding rose quartz is an effective tool for dealing with emotional stress, and a piece placed under the pillow helps improve the quality of sleep.

POLISHED

ROSE QUARTZ

FORM AND STRUCTURE
trigonal system, found in massive
deposits and also in clusters of
crystal points

COLOR
various shades of pink, from
deep rose to paler hues

GEOGRAPHICAL SOURCES
Brazil, India, Madagascar, USA

RARITY
easily obtained

HARDNESS
7

PHYSICAL/EMOTIONAL USES
supports the heart in all ways,
easing emotional stress or
regulating the physical effects
of stress, such as panic attacks

eases anxiety and fear, bringing
the warmth of compassion and
unconditional love to the system

bathes the whole body with the
healing ray of love

encourages and awakens an
appreciation of beauty

is also said to increase fertility

HEALING EFFECTS
place over the heart or on any
area of the body that needs the
support of the pink ray of
unconditional love

PERSONAL USES
wear over the heart to attract
love into your life

place in your home to create
peace and harmony

KUNZITE

Kunzite—lithium aluminum silicate—is the pink variety of a mineral called spodumene. It is unusual, forming in very defined prismatic shapes. It displays a feature called pleochroism, which means that the color is more intense when the crystal is viewed at the narrow end, looking down its length, rather than from the side. It has a tendency to splinter because it forms in parallel sheets, or flakes, that fracture in a specific direction. It has visible vertical parallel striations, or lines, on its surface caused by its structure. Kunzite was named after the mineral collector George Frederick Kunz, who first discovered it in Connecticut in 1902. It fades if exposed to sunlight, so to preserve the lovely pink color, keep it in the shade.

In crystal healing, kunzite is used to protect the heart and enable it to release emotional blocks in a gentle way. It has a soothing presence, encouraging a sense of safety and peace. It helps the heart to awaken after long periods without a relationship, when it can feel challenging to trust and open up to another person again. Kunzite is therefore a wonderful crystal to have around in the early stages of a new relationship, helping a person to welcome this new experience with a sense of joy. It also enhances healing or meditation spaces, bringing its gentle pink radiance into a room to encourage relaxation and calm.

RAW

KUNZITE

FORM AND STRUCTURE
monoclinic system, in defined
prismatic crystalline shapes,
often found in granite deposits
along with other minerals,
including quartz, feldspars,
or mica

often found alongside the green
form of spodumene (hiddenite)

COLOR
soft pink

GEOGRAPHICAL SOURCES
Brazil, Canada, USA

RARITY
fairly rare; obtain from specialist
crystal suppliers

HARDNESS
6.5–7

PHYSICAL/EMOTIONAL USES
supports the heart through
new emotional connections,
encouraging enjoyment of these
experiences

eases nervous stress, emotional
tension, and anxiety, promoting
self-confidence and joy

dissolves resistance to new ideas
and directions, enabling life to
flow and develop

HEALING EFFECTS
in healing layouts, place over the
heart to bring in the energy of
unconditional love

can also be placed over the third
eye or crown to help the influx of
spiritual energy into the system

PERSONAL USES
wear over the heart to attract new
love into your life, or to support
you during the first stages of a
new relationship

LEPIDOLITE

Lepidolite is a pink-colored mineral with a complex chemical formula: potassium lithium aluminum silicate hydroxide fluoride. It is formed when layers of lithium aluminum silicate are weakly bonded together by layers of potassium ions. It often occurs in the same locations as types of tourmaline and spodumene, and all of these minerals contain lithium. Lepidolite has a granular appearance, and it often contains sparkling particles of mica. Although prismatic-shaped lepidolite crystalline formations are sometimes available, most pieces are from mass deposits with a microcrystalline structure. Lepidolite splits in a vertical direction into thin leaves, or plates, due to its chemical makeup, forming what are sometimes called crystal books because viewed from the side the plates look like pages. Lepidolite has only become easily obtainable in the past ten years; it is now sourced principally from the USA, Africa, and Brazil.

Lepidolite's pink color links it to the heart chakra, and it releases the energy of love throughout the energy field. With its high lithium content, it is also advised for mental stress and overload, or other conditions such as hyperactivity and attention deficit disorder. It is beneficial to take a bath with a piece of lepidolite in the water in situations where there is extreme mental pressure, and anxiety or panic attacks. Lepidolite has a soothing and calming effect on the nervous system.

POLISHED

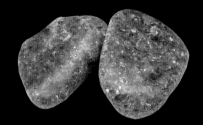

LEPIDOLITE

FORM AND STRUCTURE
monoclinic system, forming
tabular or prismatic crystals or
large microcrystalline masses,
with defined cleavage along a
vertical plane creating thin plates

COLOR
pink, also pale lilac, white

GEOGRAPHICAL SOURCES
Africa, Brazil, Russia, USA

RARITY
easily obtained

HARDNESS
2.5–3 (very soft; store with care
to avoid breakage or scratching)

PHYSICAL/EMOTIONAL USES
soothes fraught nerves and eases
extreme mental stress or tension

brings peace, calm, and a feeling
of security

helps insomnia caused by mental
overload or an inability to switch
the mind off

relieves exhaustion, whether
physical, mental, or emotional

helps relieve depression and
uplifts the mind

HEALING EFFECTS
place over the solar plexus
chakra to ease levels of physical
stress, or place over the heart
and third eye to ease mental
and emotional pressure

taking a bath with lepidolite
is a relaxing way to enjoy its
stress-relieving properties

PERSONAL USES
carry to help cope with pressures
that cause anxiety

place in your home in order to
balance the energy and create
a peaceful space

PINK TOURMALINE

Tourmaline is the name given to a large group of different but closely related minerals that appear in many different colors. Tourmaline crystals are extremely popular with mineral collectors and crystal healers because of their defined and attractive structure. The two most common types are called schorl (sodium iron aluminum borosilicate hydroxide), which is black and opaque (*see pages 208–9*), and elbaite (sodium lithium aluminum borosilicate hydroxide), which is the most common colored tourmaline sold as a gemstone. Pink tourmaline falls into the elbaite group.

All tourmalines have special properties. First, they are piezoelectric, which means that if a tourmaline crystal is heated, squeezed, or has an electrical potential applied to it, it will vibrate and show a different charge at each end, one negative and the other positive. Tourmalines are also pleochroic, which means that the color looks more intense when a stone is viewed from the top down its vertical structure, rather than from the side. Tourmalines are often found in igneous and metamorphic rocks, sometimes as inclusions in quartz and other minerals.

BLADE

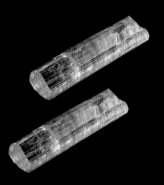

PINK TOURMALINE

FORM AND STRUCTURE
trigonal system, forming defined prismatic crystals with striations (visible lines) on the surfaces parallel to the vertical axis; found in metamorphic or igneous rocks, often embedded in quartz, which sometimes makes separation of tourmaline crystals more difficult

COLOR
pink, also green, red, brown, yellow, black, lilac, violet

GEOGRAPHICAL SOURCES
Afghanistan, Africa, Brazil, USA (pink and other colors); Madagascar, Russia, Sri Lanka (other colors)

RARITY
rare; obtain from specialist crystal suppliers

HARDNESS
7–7.5

PHYSICAL/EMOTIONAL USES
soothes the heart, eases anxiety and fraught emotions

is a useful crystal to use with children, especially those with high sensitivity or behavioral issues; also connects with the inner child in adults, the deep place that can be extremely vulnerable

soothes the physical organ of the heart by reducing stress and helps balance an irregular heartbeat

HEALING EFFECTS
place over the heart chakra to ease emotional stress, or over the solar plexus chakra to ease adrenal overload

placed on the third eye it helps to calm overactive thought patterns

PERSONAL USES
place in a bedroom or living space to create a calm and peaceful atmosphere

wear over the heart to soothe the emotions

RHODOCHROSITE

Rhodochrosite is a beautiful pink mineral with white bands, composed of manganese carbonate. The name comes from the Greek word for pink. It is sometimes found in a crystalline form in well-shaped rhombohedra, but these are rare. More commonly, rhodochrosite occurs in massive deposits, taking granular, borytroidal (grapelike), globular, or stalactitic shapes; it is often found filling cavities in the rock as well. One of the best-known sources of the mineral is Colorado, where the Sweet Home Mine produces superb specimens. The massive form of the mineral is often cut and polished into slices, spheres, or smaller pieces, as well as being carved into figurines or statues. The contrast between rose pink and white within the stone is its most attractive feature; every piece has a unique pattern.

In crystal healing, rhodochrosite is used to balance the fiery energies of the lower chakras with the unconditional love of the heart chakra. This stone helps to channel intense drives and needs into positive outcomes for higher good. It helps to stabilize the emotions when embarking on a new relationship, bringing a steadiness when feelings are running high. It also calms the heart and gives a sense of courage when facing new situations. It stimulates inner creativity and the self-confidence to make ideas take form in the physical world.

POLISHED

RAW

RHODOCHROSITE

FORM AND STRUCTURE
trigonal system, forming either defined prismatic crystals (rare) or massive deposits in various shapes, including globular or stalactitic, and sometimes filling veins or cavities in the rock

COLOR
deep rose pink banded with creamy white

GEOGRAPHICAL SOURCES
Argentina, Canada, Peru, South Africa, USA

RARITY
easily obtained

HARDNESS
3.5–4 (moderately soft, so scratches easily)

PHYSICAL/EMOTIONAL USES
strengthens and calms the heart during new life events, bringing steadiness and a sense of center

calms and rebalances the nervous system, particularly if stress is chronic or prolonged

helps recovery after an emotional trauma

dissolves fiery or deep-set emotions such as anger or jealousy, applying the healing of unconditional love to difficult situations

HEALING EFFECTS
place over the heart chakra to strengthen it and stabilize its energy, balanced by a smoky quartz between the feet to transmute negativity

place a piece in each hand to balance the energy field and give a sense of stability

PERSONAL USES
wear or carry rhodochrosite to give you self-confidence in new situations or relationships

take a bath with a stone in the water to ease emotional tension and relax the system

PINK SAPPHIRE

Although blue is the most common color associated with sapphire, pink is another example of the many colors in which this mineral is found. Sapphires—composed of aluminum oxide—are all types of corundrum, one of the hardest of known minerals apart from diamond. The white or clear variety carries the name corundrum, the red variety is ruby, and the other colors are known as types of sapphire. Trace amounts of elements, such as iron or titanium, create the variations in color; pink sapphires are colored by chromium, and if they are a deep pink, they can cost almost as much as rubies.

The word *sapphire* comes from the Hebrew *sapir*, and it is a gemstone that is mentioned many times in the Bible. Sapphires were highly prized in the ancient world, worn or carried by royalty and those of high birth in the civilizations of Egypt, Greece, and Rome.

In crystal healing, pink sapphire is a rare and beautiful stone that brings the energy of love into the physical body, encouraging playfulness, spontaneity, and the open expression of feelings. It helps dissolve old pain barriers, releasing a person into the flow of life. From a mineralogical point of view it is close to ruby in its structure—both contain chromium—but it does not have the same intensity of color or the same fiery energy. Pink sapphire is gentle and easy on the system.

POLISHED

PINK SAPPHIRE

FORM AND STRUCTURE
trigonal system, forming
prismatic, bipyramidal crystals or
rhombohedral shapes; because
of its hardness it is easily cut into
sparkling gemstones

COLOR
soft to deep pink, also blue,
yellow, green, white,
multicolored

GEOGRAPHICAL SOURCES
Africa, Australia, India, USA

RARITY
rare as a mineral specimen;
obtain from specialist suppliers

HARDNESS
9

PHYSICAL/EMOTIONAL USES
encourages new love to appear
in life, and the ability to embrace
new growth with joy and
spontaneity

eases depression and low spirits,
lightening heavy feelings of
despondency and giving hope
for the future

encourages open and loving
communication

HEALING EFFECTS
in layouts, place over the heart to
help it to open to love, or place it
over the throat to encourage the
expression of feelings

works well in combination with
other pink stones (kunzite or rose
quartz, for example) to dissolve
old levels of pain and open up to
new life

PERSONAL USES
wear to enhance your feelings
and express your love for yourself

THE SILVER RAY

Silver is the color of moonlight—bright and clear, and yet subtle. If you go out into the night when there's a full moon, it is startling how clearly everything can be seen in that light; yet the effect is soft and subdued, unlike the full glare of the sun. Silvery moonlight is just as powerful as sunlight but in more subtle ways.

As the moon always changes its shape and intensity, so it symbolizes the flowing dance from one state to another that is characteristic of silver energy. The female menstrual cycle, which is also a pattern of changes and fluctuation, is intimately linked to the moon. Research in the USA has shown that full moonlight has an effect on the pineal gland in the brain that controls hormone cycles. Animals are also affected by this light—for example, they tend to be more active during a full moon. From the Latin *luna* (moon) comes the word *lunatic*, which refers to unaccountable shifts in moods and feelings that some have attributed to the power of the moon.

The night sky

Silver also has a light-filled quality that reminds us of the stars. Native Americans believe that the lights in the sky are the campfires of the ancestors. The generations that have gone before are held in reverence, their wisdom valued. The night sky is a reminder of the continued presence of the ancestors watching over those who still live on the earth. Because modern life is increasingly lived indoors, particularly at night, we have lost touch with distant ways of life that involved meditation on or observation of changes in the heavens, perhaps while sitting around a campfire. In the warmer summer months, it is wonderful to be outdoors observing the subtle and beautiful silver energy of the night sky. Cosmic events, such as eclipses of the moon, are particularly dramatic and can be seen even without a telescope. Being under the stars reminds us that there is so much scope in the universe.

In spiritual healing, the silver ray is an influx of light, cleansing and dissolving old emotional patterns that no longer serve us. It helps to realign all the chakra centers and prepare them for the influx of new energy. A simple meditation with silver is to imagine the whole body filled with silver light that is then breathed out through the feet and into the earth. This kind of spiritual cleansing is a useful practice for those who work in caring capacities, taking on responsibility for others, as it helps shed any unwanted energy clinging to the system.

SILVER

Silver is a native element—it is found pure in nature—although it is more often found mixed with lead and copper. It has the symbol Ag, from the Latin *argentum*. Archeological excavations in Asia Minor and the islands of the Aegean Sea have revealed traces of smelting processes, showing that silver was being separated from lead as far back as 4000 BCE. Silver has been used as currency for thousands of years, too—Judas Iscariot betrayed Jesus to the soldiers of the High Priest for thirty pieces of silver—and in many languages the words for silver and money are still the same, the French *argent*, for example.

The healing properties of silver were known to the Ancient Greeks. Hippocrates, the father of modern medicine, writing around 500 BCE, suggested that silver was effective against disease. Before the arrival of modern antibiotics, silver compounds were successfully used to treat infection, and even to this day silver-coated dressings are used to prevent infection in serious burns.

Today, pure examples of native silver are extremely rare. Most silver is extracted from silver-bearing ores, such as proustite and pyrargyrite. It also occurs in veins around gold deposits. Refined commercial grade silver is 99.999 percent pure with no contamination from heavy metals or arsenic. Silver is extremely soft and can be flattened into sheets or rolled into delicate wires, making it ideal for jewelry.

SILVER INGOT

SILVER

FORM AND STRUCTURE
cubic crystal system, although rarely found as dodecahedrons, octahedrons, or cubes

sometimes found as wiry formations or nuggets, but more often as trace amounts combined with other mineral deposits, such as pyrargyrite, so requiring removal and refining

COLOR
silver-gray

GEOGRAPHICAL SOURCES
Australia, Canada, Mexico, USA

RARITY
easily sourced as jewelry; mineral specimens need to be sourced from specialist suppliers

HARDNESS
2.5–3 (very soft)

PHYSICAL/EMOTIONAL USES
is used to calm and soothe strong emotions, to help coping in the face of change, supporting the system during major upheavals in life such as the menopause

encourages intuition to be developed and trusted so that leaps in awareness can occur

strengthens the body against infection and immune dysfunction

HEALING EFFECTS
place silver nuggets or cleansed silver jewelry on the thymus chakra (over the breastbone) to support the immune system

PERSONAL USES
place a cleansed silver bracelet in water and leave it for an hour, then drink this water to purify the system

HEMATITE

Hematite is an iron ore composed of iron oxide. It is an extremely common mineral, often occurring alongside jasper, quartz, or pyrite. It has a natural silver sheen that resembles mercury. It is sometimes laid down in sedimentary deposits with alternating horizontal bands of red jasper, and this particular combination is called tiger iron. Hematite is mostly found in massive deposits, as borytroidal or kidney-shaped lumps. In rare cases, hematite is found as a circular arrangement of bladed crystals called a rosette. Because of the iron content, powdered hematite can be used to make a red pigment—and mixed with water this looks like blood (hence the name, *haima* being Greek for blood). Hematite deposits turn reddish orange as they oxidize; this accounts for the mysterious protective aspects attributed to the stone, which is seen as a powerful talisman for warriors. In the north of England, for example, there are legends about the stone preserving the blood—or life essence—of warriors fallen in battle.

Another type of hematite to look out for is specularite, which is hematite in tabular form with straight surfaces that have a silver, speckled sheen. Also, in recent times, the Chinese have created hematite magnets, made from reworked and magnetized samples, and these are fun to collect.

In crystal healing, hematite is linked to preserving strength and building self-confidence. Its silvery appearance gives it a mysterious look, and its typical rounded, lumpy shapes mean it resembles parts of the human body, such as the kidneys and the brain.

GRANULAR CLUSTER

HEMATITE

FORM AND STRUCTURE
trigonal system, forming borytroidal, massive, or kidney-shaped examples; sometimes also found in sedimentary deposits as small grains; more rarely, as a circular cluster of tabular crystals called a rosette

COLOR
silver-gray, steel gray

GEOGRAPHICAL SOURCES
Australia, Brazil, Britain, Mexico

RARITY
easily obtained

HARDNESS
5–6

PHYSICAL/EMOTIONAL USES
cleanses the blood and supports the circulation

warms and energizes the system, giving a sense of strength

its silvery color can open the mind to more sensitive perception of the spirit world

grounds and centers the physical body in the present moment

acts as a protective stone, shielding the wearer from negativity

HEALING EFFECTS
place over the lower abdomen or between the feet to ground and center the physical body, along with a Herkimer diamond over the crown chakra to bring spiritual energy into physical manifestation

PERSONAL USES
wear or carry as a protective stone in order to build your self-confidence

THE BLACK RAY

Black is absorbent; it draws all other colors into itself. It is a powerful and necessary part of the color palette, not to be ignored. In Peruvian and Native American traditions, darkness is recognized as the place of initiation and change; it fosters the emergence of a personal inner power that comes from confronting fear. In the teachings of these traditions, the self we think we are in everyday life is merely a set of beliefs we have formed, largely because of our emotional reactions since childhood.

This does not reflect who we really are, and to find out the truth about ourselves we have to take a journey into the dark, let go of all those beliefs, and find the deep essence of ourselves. This is the concept of shamanic journeying, and if undertaken with the right support it can completely transform the way life is appreciated and lived.

A powerful illustration of the relationship between black and white is the yin-yang symbol. It originates in the Far East and is linked to the philosophy of traditional Chinese medicine, which makes use of different types of energy. Yin energy is dark, inner, passive, cool, and receptive; yang energy is light, outer, active, hot, and dynamic. Both are needed to keep everything in balance.

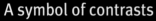

A symbol of contrasts

The swirling pattern of the symbol allows an interlocking of the two sides, colored white and black. This symbol is infinite, it is not linear; as you observe it, the eye naturally travels in a circular or spiral pattern. This is symbolic of flow between one state and another: breathing in and breathing out, being active and being still, feeling heat and feeling cold. These contrasts are vital perspectives on the ever-changing nature of life everywhere. Within the yin-yang symbol is a very powerful illustration of transformation: in each side of the image there is a small contrasting dot. What this means is that within light, there is always the possibility of darkness, and within darkness there is always the possibility of light. Meditation on this concept reveals deep inner truths about life and its meaning. Too much emphasis on light denies the darkness its role and its power, and this is often a problem with a lot of New Age philosophy. Both dark and light, yin and yang, are needed—even out in space. Stars are born, radiate brilliance, explode, and sink into black holes that are the creative cauldrons of stellar rebirth.

BLACK OBSIDIAN

Obsidian is volcanic glass, composed of silicon dioxide (quartz) and many impurities that allow it to take on different colors and shades. It forms when volcanic lava comes into contact with water. Often the lava forms into a lake as it cools rapidly in the presence of the water, producing a glassy effect in the resulting rocks. Black obsidian is colored by iron and magnesium. Sometimes small, smooth, rounded pieces can be found, which are called Apache Tears. The indigenous tribes of North America used pieces of obsidian to make arrowheads, spears, and knives. Obsidian can be cut and shaped by striking it at the correct angle, and this creates a razor-sharp edge. Examples of obsidian tools have been found in archeological investigations in Arizona dating back at least 10,000 years.

Because of its glassy consistency, black obsidian occurs in many interesting shapes, and may have a translucent appearance as well. In crystal healing, this is a stone of mystery, of connection to deeper, unknown aspects of the self. It is a good companion to carry during the period of a healing process, especially one that is unfolding over time, perhaps through several sessions with a therapist. It is a powerful psychic protector. Burying it in the ground for a night is a good way to cleanse it and enable the stone to recharge itself with protective energy.

POLISHED

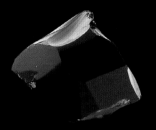

BLACK OBSIDIAN

FORM AND STRUCTURE
silicon dioxide volcanic glass in
an amorphous state (a structure
with no inner geometric patterns)

COLOR
black, also dark brown and
dark green

GEOGRAPHICAL SOURCES
Britain, Italy, Mexico, USA

RARITY
easily obtained

HARDNESS
5–5.5 (much softer than quartz)

PHYSICAL/EMOTIONAL USES
a powerful psychic protection
stone, helping to deflect needy or
draining energy being projected
by other people

stimulates the root chakra,
drawing in the earth's energy
to strengthen and stabilize
the system

brings to the surface things
that might have been otherwise
hidden, making it a useful ally
during a healing process

cuts through deception, bringing
clarity and focus

HEALING EFFECTS
in healing layouts, place between
the feet or a small piece in each
hand to ground and strengthen
the energy field around the body

PERSONAL USES
place in the home to absorb
excess electromagnetic energy

wear or carry as a stone of
protection

BLACK TOURMALINE

Black tourmaline—also called schorl—is composed of sodium iron aluminum borosilicate hydroxide oxide. This complex chemical formula creates crystals with pointed terminations made up of several facets at different angles. The flat sides of the crystal may show striations, or lines, running in parallel to the vertical axis of the stone. Iron is the main source of the black color of this type of tourmaline. It can be found in a wide variety of sizes, from small pieces to large heavy points. Like all tourmalines, this crystal is piezoelectric, meaning that heating or squeezing a crystal will create a different electrical charge at each end of the stone, one positive, one negative. Tourmaline crystals generally form in igneous or metamorphic rocks.

Black tourmaline is an especially useful crystal to keep in the home or to carry around for personal protection from negative energies, be they from other people or from the environment—electromagnetic waves, for example. The piezoelectric property of black tourmaline—meaning it can convert mechanical energy into electric energy—encourages balance of the whole body's energetic framework. It is important to maintain this balance because of the many levels of environmental stress that affect human beings in the modern world. In crystal healing, black tourmaline is one of the most popular of all the cleansing and grounding stones, used to counteract negativity and to detoxify the system.

RAW

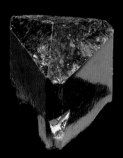

BLACK TOURMALINE

FORM AND STRUCTURE
trigonal crystal system, forming prismatic crystals with terminations (pointed ends with facets) and vertical striations

COLOR
black, also pink (*see pages 192–93*), watermelon (*see pages 252–53*)

GEOGRAPHICAL SOURCES
Brazil, Nepal, Pakistan, USA

RARITY
easily obtained

HARDNESS
7–7.5

PHYSICAL/EMOTIONAL USES
enhances the psychic protection of the body and energy system, acting as a shield against negativity

cleanses, purifies, and detoxifies the body and the energy field that surrounds it

realigns the chakras and creates a clear channel between the earth (root chakra) and heaven (crown chakra)

HEALING EFFECTS
in layouts, place black tourmaline and hematite between the feet to anchor and ground a person's energy during healing

place an amethyst or a Herkimer diamond above the head to open the crown chakra to spiritual frequencies

PERSONAL USES
wear or carry to help you stay grounded and focused on the present moment and to counter electromagnetic stress

JET

Jet—or lignite—is a special geological material. It is not considered to be a mineral in the true sense of the word; it is, in fact, a mineraloid. This is because, although it resembles a mineral, it is organic in origin, having been formed from waterlogged wood trapped under layers of mud, where the carbon has become compressed over time. Essentially, it is a hardened form of coal, and it comes in two forms: "hard," where the carbon compression happened in salt water, and "soft," where this process took place in fresh water. Like amber—another mineraloid—jet will take on a static electrical charge if rubbed. It is not cold to the touch like glass, and it has a smooth feel to it. Sometimes pieces contain patterns similar to tree rings, showing jet's organic origins.

Jet can be carved into figurines, beads, and other types of jewelry. Examples dating back thousands of years have been found in excavations of the Celtic settlement in Hallstatt, Austria. In Roman times, jet jewelry was also produced in the city of York and other parts of modern northeast England. The town of Whitby, in North Yorkshire, is a famous source of jet; in the nineteenth century, Queen Victoria wore Whitby jet jewelry as part of her mourning dress after the death of her husband Prince Albert.

In crystal healing, jet is used to cleanse and purify the physical and energetic aspects of the human body. It has a gentler energy than black tourmaline or obsidian, and it offers protection and nurturing.

POLISHED

JET

FORM AND STRUCTURE
compressed coal, forming in layers of sediment; available in chunks of varied sizes

COLOR
black

GEOGRAPHICAL SOURCES
Britain, France, Germany, Poland, Russia, Spain, USA

RARITY
easily obtained

HARDNESS
3–4

PHYSICAL/EMOTIONAL USES
surrounds the body with protection and helps neutralize negativity in a gentle way

cleanses and detoxifies the chakras, and also supports the activity of the kidneys, which dispose of physical toxins

helps with the recall of past lives and the healing of past issues persisting into this life

helps improve memory

HEALING EFFECTS
in healing layouts, place over any areas that are congested, or place a piece in each hand to assist with cleansing the system

works well in tandem with amber placed over the sacral or solar plexus chakras to replenish the system

PERSONAL USES
wear or carry as personal protection and talisman

BLACK ONYX

Onyx is a banded variety of chalcedony, a form of quartz. It is distinctive because it has parallel bands within its structure. Natural black onyx is not uniformly black; it has contrasting stripes of gray, white, and black. These are all caused by different levels of impurities within the quartz. Onyx can be found in many colors, including white, tan, and brown. It is a stone that has been used for centuries in carving, notably for signet rings or cameos with intricate designs. Examples of pure black onyx are likely to be made from stained gray or brown stones. This practice is not new; it has gone on for centuries, and involves introducing color into a stone to improve uniformity.

Naturally banded pieces of black onyx are preferable for healing, simply because they have not been artificially colored. The gray, white, and black bands are symbolic of the interconnectedness of black and white, as previously discussed with the yin-yang symbol. Black and white are at opposite ends of the spectrum, and gray is the color they form when mixed together. These three colors are like different levels of existence: white is the light of everyday life, black the realm of sleep and the unconscious, and gray the color of predawn or dusk, twilight moments when all is still. In fairy tales, dusk and dawn are magical times. Banded black onyx is a reminder of the interconnectedness of these different levels of existence.

POLISHED

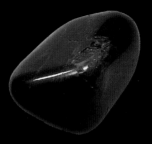

BLACK ONYX

FORM AND STRUCTURE
microcrystalline chalcedony quartz with parallel bands

COLOR
black stripes mixed with gray and white

GEOGRAPHICAL SOURCES
Brazil, Germany, Mexico, USA

RARITY
common and easily obtained

HARDNESS
7

PHYSICAL/EMOTIONAL USES
black onyx in its banded form helps a person balance the demands of everyday life with the need to find rest and new inspiration

helps restore the system if it is overloaded because of excess pressure

encourages inner stillness and helps focus during meditation

energizes the feet and legs, working on the root chakra

HEALING EFFECTS
place between the feet or hold a piece in each hand or over the lower abdomen at root-chakra level to ground and stabilize the physical and energetic systems

PERSONAL USES
wear or carry to improve the balance between work and personal life, especially if there is too much emphasis on work

THE WHITE RAY

Pure white is the color of snowflakes, each one of which is unique and perfect in geometric symmetry. Looking at a patch of virgin snow in sunlight reveals more than simple white; the eye may also detect sparkles of rainbow refractions from frozen crystalline forms within the snowdrifts. White is not uniform. It is brilliant and sparkling, and it has a clarifying radiance that lifts the spirits. It is a combination of all the colors of the visible light spectrum.

214 If a circle of thin cardboard is painted with the entire rainbow of colors, perforated with a toothpick and spun quickly enough, all the colors will merge and disappear into white. This is the energy that is used in healing to cleanse, purify, and replenish the energy blueprint of the human being. In Jin Shin Jyutsu, an ancient Japanese art that demonstrates how energy relates to physical form, the concept of the energy framework of the human body is crucial. According to this teaching, this framework exists before the physical body takes form and it is a blueprint that can be restored whenever energy is applied to the body. No matter how the physical body has been affected by circumstances, the blueprint helps to realign and retune the system to original levels of energy, which in turn brings healing to the body.

A symbol of transformation

White and clear crystals are one of the biggest groups of stones, and are popular as healing tools. The ultimate example in this group has to be diamond. In its rough state it looks opaque and dull, but when it is faceted and polished all kinds of fiery brilliant rainbow reflections are released. Within a diamond, all imaginable colors are reflected, all held in one structure, in unity. The diamond is a powerful symbol of the transformation of the human being from a physical state to a light-filled state. This is the goal of many spiritual traditions, illustrated by the phrase: "Feet on the ground, head in heaven." Human beings have the capacity to be bridges between the spiritual and physical worlds, and in so doing support and work in harmony with the earth. In meditation, imagine a diamond sparkling over the top of the head, drawing white light into the body, then pass it out through the feet. This cleanses your individual energy system and passes energy into the earth. It is a simple and wonderful way to interact with the planet.

WHITE MOONSTONE

Moonstone is a descriptive name for a mineral known as oligoclase, which is a type of feldspar composed of potassium aluminum silicate. It can be found in a number of colors, but all examples demonstrate a particular sheen known as labradorescence, an effect also found in labradorite (*see pages 124–25*) and blue moonstone (*see pages 142–43*). Labradorescence is a light effect caused by parallel hairlike structures laid down within the crystal matrix, which reflect light back and forth, creating different-colored effects, including blue, yellow, silvery gray, and white. There is a history of its use in many cultures, from India and the Far East to the Celtic peoples of Europe. It has been polished and set in silver or gold for centuries, and some fine examples of ancient moonstone jewelry can be seen in the British Museum.

This type of moonstone has a soft white sheen and of all these stones it looks most like the moon itself; it particularly recalls the shimmering glow of the full moon in a clear sky. Hindu tradition links this stone to the divine feminine principle, and it is a symbol of love. The ever-changing, almost fluid appearance of the stone recalls the fluctuating changes of moods and hormones throughout the menstrual cycle. When the moon is full, the creative power of the feminine is at its height. White moonstone helps to release this creative impulse and sustain it.

TUMBLESTONE

WHITE MOONSTONE

FORM AND STRUCTURE
feldspar in a triclinic system, forming in large microcrystalline masses with lamellar intergrowths forming a sheen in certain lights

COLOR
white, cream, or yellow base with white, blue, gold, or gray sheens

GEOGRAPHICAL SOURCES
Australia, India, Myanmar, Sri Lanka, USA

RARITY
easily obtained; lower-grade tumblestones do not generally show much of a sheen

HARDNESS
6.5

PHYSICAL/EMOTIONAL USES
helps balance the hormones during the menstrual cycle, and also assists during times of major hormonal change such as puberty or menopause

may also help hormonal issues linked to infertility

its effects are not limited to women, as it helps men become more receptive and in touch with inner feelings

increases creativity and enhances intuition

HEALING EFFECTS
in healing layouts, place over the heart and third-eye chakras, and over the lower abdomen (root chakra) to balance the hormones and ease emotional stress

PERSONAL USES
wear over the heart to soothe the emotions, balance feminine energy, and harmonize the hormones

SELENITE

Selenite is hydrated calcium sulfate, a type of gypsum. This is an extremely common mineral that forms in sedimentary environments— around hot springs, for example. It is found in massive beds, usually crystallized out of highly saline water. Selenite is the colorless and transparent form of gypsum. It has a pearl-like sheen and a glow like moonlight. The name selenite is from the Greek moon goddess Selene. The crystal forms clear blades with striations—visible lines—along its vertical plane, and some pieces of selenite contain air or water bubbles. Large pieces of transparent selenite are common, as are smaller crystal blades. Selenite has some interesting qualities: It is extremely soft, so some pieces will actually bend slightly; it also acts as an insulator, so it feels warmer to the touch than other clear crystals.

In crystal healing, blades of selenite make natural wands with a powerful action. They act like conductors of energy and can be felt acting on the body like a rushing wave, clearing away physical and emotional toxins and negative mental patterns. If other crystals are fixed to selenite in wands, it makes their action even more powerful. For this reason, it is best to leave specialized energy work with selenite to qualified crystal healers. However, selenite can still be enjoyed by placing a piece in the home, where it will maintain positive energy.

BLADE

SELENITE

FORM AND STRUCTURE
transparent or translucent crystals in a monoclinic system, forming in flat-faced tabular, bladed, or blocked formations

twin crystals are common, sometimes in a shape called swallow-tail

COLOR
clear with a pearl-like sheen on flat surfaces

GEOGRAPHICAL SOURCES
Italy, Mexico, USA, but common around the world

RARITY
easily obtained

HARDNESS
2 (can be scratched with a fingernail)

PHYSICAL/EMOTIONAL USES
is a conductor of energy, and its cleansing effects can be strongly felt on the system

allows a flow of high frequency energy to permeate all the chakra centers, aligning them all with spiritual purpose

resonates powerfully with the crown chakra, enabling enhanced awareness of enlightenment

HEALING EFFECTS
place a selenite blade over the crown of the head to open the crown chakra—if you are not experienced in crystal healing, only do this for a few moments; ground the body by placing smoky quartz between the feet

PERSONAL USES
place selenite in the home or in a meditation space to clear the energy

DIAMOND

Diamond—which is made of pure carbon—is the hardest substance found in nature, measuring 10 on the Mohs Scale of hardness. It is used as a benchmark against which all other crystals and gemstones are measured. Diamond is actually four times harder than corundrum (sapphire and ruby), which is the next hardest mineral. However, diamond also has an interesting feature in that it has four potential directions of cleavage, so if it is hit in any of these directions, it will split. Skilled jewelers have to cut diamonds so that these angles do not present themselves when the stones are worn in settings. Diamond is also a powerful conductor of heat—even better than silver—and its melting point is at 6,416 degrees Fahrenheit (3,547 degrees Celsius). Diamond shares the same chemistry as graphite (used in pencil lead), which is also pure carbon. However, the two have a different structure, so diamond is transparent while graphite is opaque.

There are many famous diamonds in the world. A notable example is the Koh-i-Noor diamond from India, which, after possibly thousands of years of being fought over as a spoil of war, found its way into the British Crown in 1877. Weighing in at 105 metric carats (21.6 grams), this was the largest polished diamond in the world until the twentieth century.

Diamonds are rare as readily available mineral specimens because of their use in jewelry. Small, double-terminated, and highly faceted crystals called Herkimer diamonds are available; however, these are not, in fact, diamonds, but a type of clear quartz that share some of its properties (*see pages 228–29*).

FACETED

DIAMOND

FORM AND STRUCTURE
carbon in an isometric system, often forming cuboid or octahedron-shaped crystals, and often associated with such minerals as kimberlite

COLOR
in its rough state, cream or yellowish and opaque, only becoming brilliant when cut; also blue, black, red, green

GEOGRAPHICAL SOURCES
Australia, Brazil, India, Russia, South Africa

RARITY
rare as mineral specimen; obtain from specialist suppliers

HARDNESS
10

PHYSICAL/EMOTIONAL USES
carries the highest healing frequency, opening all chakras and expanding the energy field, or aura, around the body

encourages the full awareness of spiritual energy within the physical body

recalls a person's true destiny and purpose

HEALING EFFECTS
four uncut diamonds—one at the head, one between the feet, and one by each hand—creates a light grid around the body that strengthens the aura

PERSONAL USES
wearing a diamond is a powerful amplifier of spiritual energy

DANBURITE

Danburite is a calcium borosilicate mineral that is becoming popular. Its crystalline shape resembles topaz; it forms prismatic crystals with a diamond shape across the horizontal axis and clear striations running along the vertical axis. It can be found in large examples of up to a foot (30 centimeters) long, or in smaller blades. The wedgelike shape of these crystals helps to differentiate them from quartz, which forms in hexagonal crystals. Polished tumblestones of danburite have a more watery look than quartz and danburite is also considerably lighter. The name comes from the stone's original locality, the city of Danbury in Connecticut; however, the crystal is now sourced in other countries, including Mexico.

Danburite is an example of a "new" crystal that has only recently come into prominence. Many different varieties of clear crystals are now appearing, and in healing circles this is seen as the earth providing tools to help humankind in its evolution toward light. In the case of danburite, the energetic vibration is gentle and nurturing; this is a crystal to keep or use to assist a patient unfolding of consciousness. Some people are ready for rapid and powerful shifts; others need to change more slowly and steadily. Danburite's vibration is loving and soft, and it establishes communication with the angelic realms, as well as personal guides in the realm of spirit.

TUMBLESTONE

DANBURITE

FORM AND STRUCTURE
orthorhombic crystal system, producing prismatic crystals with a diamond-shaped cross section

the termination is usually steeply slanted, which produces a wedgelike look

COLOR
usually white or clear, also pink, yellow, brown

GEOGRAPHICAL SOURCES
Japan, Mexico, Myanmar, Switzerland, USA

RARITY
obtain from specialist suppliers

HARDNESS
7–7.5

PHYSICAL/EMOTIONAL USES
opens awareness of the angelic frequencies and angel guides, bringing a gentle energy of healing to the whole system

balances the heart chakra with the crown chakra

soothes the nervous system and eases nervous tension or stress

enhances creativity and intuition

HEALING EFFECTS
in healing layouts, place one piece of danburite over the heart and another over the crown chakra to balance the energies of these centers and soothe the nervous system; add a piece of smoky quartz between the feet to remain grounded

PERSONAL USES
make a gem remedy (*see pages 284–85*) with danburite and take in water to feel relief from stress

carry danburite in order to enhance your creativity

AZEZTULITE

Azeztulite is not a proper mineral name, it is a trade name given to a type of quartz from North Carolina, first discovered in 1970. Azeztulite crystals tend to be expensive for what appear to be roughly formed, small pieces of opaque quartz with an irregular shape. The reason for this is that they are in limited supply from the North Carolina source, and this increases the price. Interest in Azeztulite has grown because it is seen as a New Age crystal, holding powerful energies and enabling high levels of personal transformation in line with changes to the planet. Crystal healers in particular seek out this stone wherever it can be found to use as a healing tool.

Azeztulite is a name that the first therapists who worked with the mineral received in psychic transmissions. This crystal resonates at such a high frequency that it is said never to need cleansing, it simply transforms and transmutes negativity. This is a powerful tool, and it is therefore a good idea to meditate or work with it in the presence of a qualified crystal healer. This is because Azeztulite has a powerful releasing effect within the system, and sometimes help is needed to understand the energetic fallout from such reactions, which can be emotional and linked to deep memories. Clearing away these patterns helps to prepare the physical, mental, emotional, and spiritual aspects of the self for an influx of light from the higher realms.

RAW

AZEZTULITE

FORM AND STRUCTURE
trigonal crystal system, forming quartz in small irregularly shaped pieces, some darker and opaque, and some transparent

COLOR
clear, white, cream

GEOGRAPHICAL SOURCES
USA

RARITY
very rare; obtain from specialist suppliers

HARDNESS
7

PHYSICAL/EMOTIONAL USES
helps to clear chronic patterns of disease from the system, such as persistent low immunity

clears old mental patterns and opens new levels of perception and understanding

helps to gather courage for a new start in life

prepares the system for an influx of higher vibratory frequencies of light

HEALING EFFECTS
in healing layouts, a piece of Azeztulite can be placed over the heart chakra to open it and release old emotional patterns

placed over the third eye, it assists the opening of psychic perception

PERSONAL USES
because it carries a powerful vibration, Azeztulite is best worked with in conjunction with a crystal-healing professional

PHENACITE

Phenacite—sometimes spelled phenakite—is composed of beryllium silicate. It is often found in pockets of pegmatite with other gemstones in the beryl group, such as topaz or emerald, and it also occurs with smoky quartz or chrysoberyl. Its name comes from the Greek word *phenakistes*, meaning cheat, because it is often mistaken for other stones, most commonly quartz and topaz. It tends to form in short hexagonal prisms, sometimes in a twinned form, and often has striations, or lines, along the vertical axis. Phenacite crystals display good levels of hardness, so they can be cut and faceted as gemstones. Often the crystals are small and consequently require careful handling. Examples of phenacite from different countries are said to hold slightly different vibrations, so it is useful to ask a supplier for the geographical source of a stone. Brazilian phenacite, for example, has a clearing vibration, whereas Russian is said to have a more stimulating effect.

Phenacite is another example of the clear stones that are gaining prominence as powerful cleansing and spiritual-transformational tools. It is considered to resonate with the third-eye chakra, opening the doors to higher levels of psychic perception. It also enables individuals to become conscious of their soul purpose, and to bring the energy of such perceptions into the realm of everyday life. Opening to such awareness often generates changes in life; to deal with this, it is beneficial to obtain support from an experienced crystal healer in order to understand and assimilate the lessons learned.

RAW

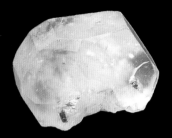

PHENACITE

FORM AND STRUCTURE
trigonal system, forming transparent or translucent crystals in a rhombohedral or prismatic shape, often in small or short pieces

COLOR
usually colorless or whitish, but sometimes yellow, pink, brown

GEOGRAPHICAL SOURCES
Brazil, Madagascar, Norway, Russia, USA

RARITY
rare; obtain from specialist suppliers

HARDNESS
7.5–8

PHYSICAL/EMOTIONAL USES
clears and opens the third-eye chakra, amplifying psychic awareness

activates the energy blueprint of the body, enabling healing to transfer from the energetic level to the physical body

supports brain function, and its energy may be supportive to people with dementia

activates intuition and inner knowing

HEALING EFFECTS
in healing layouts, place over the third-eye chakra; this is best done with a crystal healer as it is a powerful transformational crystal and the system needs to be prepared to receive its energy

PERSONAL USES
phenacite's energy is best experienced through healing from a qualified therapist

CLEAR QUARTZ

Clear quartz is probably the most popular crystal of all for healing, and a common choice when starting a new collection. To many people, clear quartz is synonymous with the word *crystal*; it is the ultimate image of a beautiful mineral. Quartz is silicon dioxide and is found in abundance all over the world. It can take many forms, from large six-sided crystals with defined faceted ends called terminations to intricate clusters of fine needlelike points. Quartz is often found in granite deposits or in sedimentary rocks such as sandstone. Other crystals—including rose quartz, amethyst, and citrine—are actually clear quartz that has been colored by impurities.

Most often clear quartz is, as its name suggests, transparent; however, an opaque form does occur—called snow quartz or milky quartz—where opacity is caused by tiny bubbles trapped within the structure. Many clear-quartz crystals have some traces of air or water bubbles in them, creating areas of cloudiness. In fact, quartz crystals occur in an amazing diversity of shapes, sizes and levels of transparency, containing bubbles or other faults that catch the light and create rainbow reflections. These features make clear quartz one of the most beautiful crystals to collect.

In crystal healing, clear quartz is used to cleanse, focus, and amplify energy levels in the body. Clear quartz crystals can be programed, meaning they can reflect the intentions of those who work with them, amplifying the potential for those intentions to take form. To benefit the earth and humanity, it is important to use them for the highest good.

SNOW QUARTZ
CLUSTER

SNOW QUARTZ
POLISHED

FORM AND STRUCTURE
hexagonal structure, creating
defined six-sided crystals with
flat surfaces and defined faceted
termination points

COLOR
transparent, sometimes with
cloudy white areas

GEOGRAPHICAL SOURCES
Brazil, Madagascar, USA, but can
be found worldwide

RARITY
easily obtained

HARDNESS
7

PHYSICAL/EMOTIONAL USES
clear quartz (snow quartz has
similar effects but is gentler)
is a cleansing tool, clearing
impurities and psychic debris
from the energy field that
permeates the human body

realigns and refocuses the
body, mind, and spirit toward
clear goals

increases power of concentration
and focus

amplifies whatever thoughts
have been programed into it

HEALING EFFECTS
in layouts, place points over the
crown, solar plexus, and root
chakras to clear the system, or
one in each hand to balance the
left and right sides of the body

PERSONAL USES
place in your home to cleanse
the energy and enhance peace
and clarity

CLEAR QUARTZ VARIATIONS

On the following eight pages there are descriptions of some of the most important specialist varieties of clear quartz formations. These are quite specific, and all have particular meanings and uses in crystal healing. If you are drawn to a particular shape or type of specialist quartz, it means that this tool has importance for you at the time of choosing. Meditation with a specialist piece is crucial so that the significance of the crystal, and methods that may be helpful to you in using it, can become clear. Professional crystal healers also use these tools, and it can be interesting and useful to consult them for advice.

All these variations of clear quartz make interesting and beautiful additions to any crystal collection; however, it is also important to recognize them as powerful healing tools. Experienced crystal healers believe that often crystals choose their keepers, and not the other way around. If an unusual piece comes to you, then sit with it, meditate with it, or discuss it with a qualified therapist. Listen to the information or guidance you receive. It could be that a particular crystal tool is telling you that you have a special task to perform or a service to give to the earth and to other people.

DOUBLE-TERMINATED

These examples of clear quartz grow into a natural point, or termination, at each end of the crystal. Normally clear quartz has only one pointed end, which is able to channel and direct energy; double-terminated pieces can do this from either end. Worn as a pendant, a double-terminated piece helps to channel energy up and down the chakra system, enhancing its flow and direction. It also helps to bridge spirit and matter, bringing both together in harmony. Special, brilliantly clear small double-terminated quartz examples from New York State are called Herkimer diamonds.

RUTILATED CLEAR QUARTZ

This is clear quartz containing golden-colored strands of rutile, a mineral composed of titanium oxide. The strands look almost like golden hair trapped within the structure of the quartz itself, and they show up dramatically in the light. These strands are seen as energy conductors, making rutilated quartz even more powerful as a cleansing tool as well as an excellent assistant in psychic expansion. It allows spiritual energy to be conducted through the system, aligning all the chakra centers. Small pieces can be placed at the crown, over the heart, between the feet, or in each hand to strengthen the energy field.

ELESTIAL

Often found as smoky quartz, and sometimes containing sparkling particles of mica, elestial quartz has a stepped appearance and many facets. It is a powerful stone for meditation, bringing clarity and focus to overwhelming life situations, and perhaps highlighting issues in past lives, as well as this one, which are ready to be transformed through love. Although it has many facets, it also has a unified shape, as a reminder that all apparently fragmented aspects of life can be brought into wholeness and integration.

PHANTOM

This type of clear quartz has a second, cloudy crystal outline that is faintly visible within the structure of the main piece. This is actually a sign of a phase of previous growth within the crystal. In healing, phantom quartz helps to shed light on aspects of life that are hidden or seem mysterious, particularly things that relate to the past. It can also signify life events that never came to fruition. Phantom quartz enables the healing of past issues by bringing them into the light of the present moment, where they can be clearly seen and understood. Carry it or meditate with it to feel the effects.

LASER WAND

A laser wand is a long thin blade of clear quartz with a defined tapered point and small facets. This type of crystal serves as a light tool, able to focus and direct light from its sharpened end. This projected light can then be used to remove density, blockages, or areas of stagnation from the energy field. It is a precise crystal and needs to be used with awareness and a clear focus. If you are new to crystals, it is best to experience the energy of crystal laser wands via the hands of a professional crystal healer.

233

MANIFESTATION

Manifestation quartz is an amazing phenomenon, in which one or more smaller crystals are clearly visible within a larger crystal. Unlike the cloudiness in phantom quartz, these are transparent and are fully formed crystals in themselves. The larger piece is like the mother of the smaller crystals. Manifestation quartz is a precious symbol of the potential that exists within each one of us to make our dreams come true. Meditation with a piece of this crystal brings strong creative visions and shows the user how to bring these into the world of physical reality.

SELF-HEALED

A self-healed quartz piece has broken at some point during its formation and regrown to heal the break. The area that has reformed is usually cloudy, and the crystal has changed axis, or direction, in its new pattern. Self-healed quartz is a symbol of the power of healing.

If there is a deep wound in the skin, for example, it leaves a scar, a sign of its presence. However, the new scar tissue that forms is stronger than what was there before. Self-healed quartz shows that life's wounds can be great teachers, enabling renewal.

GENERATOR

Generator quartz is a rare formation with six perfectly equal sides and a point made up of six identically angled facets. Perfect naturally occurring symmetry is very unusual in the mineral kingdom because of the constantly changing relationships between chemistry, pressure, and temperature during the crystal-formation process. This is why most crystals are asymmetrical. Generator quartz, as its name implies, is like a battery that helps to recharge depleted people and spaces. It is also a good focus for group meditation and prayer because it amplifies spiritual energy.

SOUL-MATE

This is a beautiful form of twinned quartz, where two crystals grow together from the base but separate to form two distinct points. This is called a soul-mate or twin formation. It is symbolic of a relationship in its truest sense, where there is a base in union but the two energies involved still maintain their unique identities. A soul mate is another person with whom there is deep and loving interaction, based on spiritual understanding. Placing a soul-mate crystal in the home makes it possible for such a relationship to manifest in life.

235

CHANNELING

A channeling piece of quartz is distinctive because it presents one main facet at its point with seven clear sides and a small opposite facet with just three sides. The numbers seven and three are significant: seven is the number of the chakras, the energy centers making up the human aura; three is the number of the dimensions we live in as physical reality—height, width, and depth. Channeling quartz draws in energy through the seven chakras and allows it to ground in the earth. This is a powerful form of quartz, and it is best to consult a crystal healer about where to place it and how to use it.

CATHEDRAL

Cathedral quartz is one of the most beautiful and unusual of the quartz variations. It is given that name because it has a stepped structure that resembles the towers and buttresses of Gothic architecture. Crystal healers regard it as the key to levels of knowledge and awareness beyond the physical world, sometimes called the akashic record. This is said to be a field of information that has been accessed throughout time by seekers, mystics, and psychics, and which holds answers to deep mysteries. Holding cathedral quartz may bring flashes of awareness, memory, or déjà vu that shed light on inner issues.

RECORD-KEEPER

Record-keeper quartz has unusual triangular or chevron-shaped marks on one or more surfaces of its long sides. These crystals are quite rare, and the pattern of marks is unique to each piece. Crystal healers regard record-keeper quartz as another repository of soul memory, possibly linking to past lives that can be recalled. This type of quartz is deeply harmonizing to body, mind, and spirit, and it helps develop the power of memory. Placed over the third eye, it can activate detailed recollections of places and people beyond this lifetime.

ABUNDANCE

Abundance quartz takes the form of a cluster of many small crystals with one or possibly two larger points emerging from it. This complex type of cluster holds crystals in many different angles, acting like a receiver to bring in light. Abundance means trusting in the universe to provide all that is needed in life, not just money or wealth, but also people, experiences, and opportunities for growth. Placing an abundance crystal in the home helps to encourage new directions and a positive flow of energy to come in. Meditate with the crystal and focus on receiving with gratitude.

237

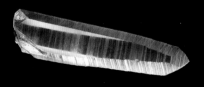

LEMURIAN-SEED

This is an unusual form of quartz that has only recently become widely available. It has a very faint pinkish tinge, and along its sides it has horizontally aligned striations (lines), usually all the way up to the facets at the tip. Its energy feels very different from regular quartz; when held in the hand there is a tingling warmth.

Crystal healers link this type of quartz to Lemuria, a hypothetical lost land that predates the story of the Flood in the Bible, where the people had a close and instinctive relationship with the earth. Lemurian-seed quartz awakens the soul of the person working with it to a deep consciousness of the planet and its needs.

Crystals Gallery

The following pages show a wonderful collection of images of many kinds of crystals in all their colors and beauty. These pages are a visual treasure chest to be enjoyed and explored.

Looking at these images may inspire you in different ways. For example, you may be drawn to crystals of a certain color and feel that you would like to have a particular stone with you as a result. If you select a stone in this way, don't forget to check its individual profile for all its qualities and its color healing links. On the other hand, you may be looking for a particular

type of crystal and this gallery will show you examples so that you can recognize specimens in stores. You may also be given a crystal and wish to check what kind it is—again, the gallery is a good place to start looking so you can identify the stone and learn how it can help you.

The world of crystals is fascinating, diverse and magical. Their physical qualities—some geometric, some metallic, some transparent, some opaque, some filled with reflections and some dark and mysterious— are all displayed here for you to see.

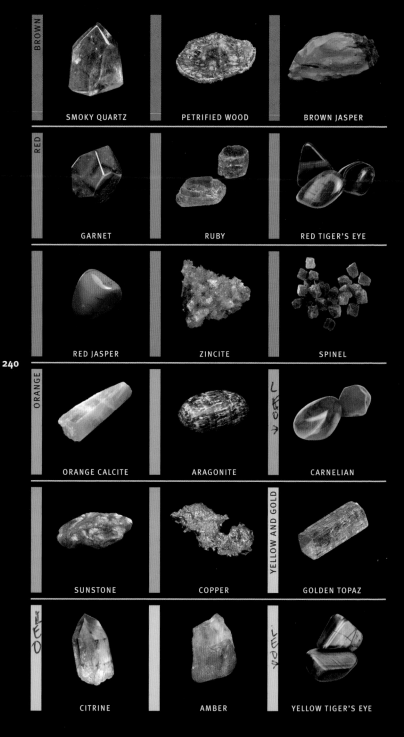

BROWN

SMOKY QUARTZ

PETRIFIED WOOD

BROWN JASPER

RED

GARNET

RUBY

RED TIGER'S EYE

RED JASPER

ZINCITE

SPINEL

ORANGE

ORANGE CALCITE

ARAGONITE

CARNELIAN

SUNSTONE

COPPER

YELLOW AND GOLD

GOLDEN TOPAZ

CITRINE

AMBER

YELLOW TIGER'S EYE

YELLOW JASPER

SULFUR

CHRYSOBERYL

PYRITE

GOLD

GOLDEN GREEN

PERIDOT

CHROME DIOPSIDE

SERPENTINE

PALE GREEN

APOPHYLITE

PREHNITE

GREEN CALCITE

GREEN

EMERALD

AVENTURINE

GREEN FLUORITE

JADE

MALACHITE

DARK GREEN

MOLDAVITE

MOSS AGATE

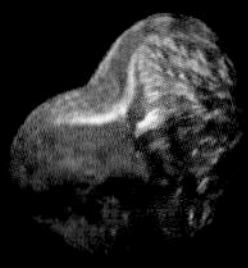

SERAPHINITE

BLUE-GREEN

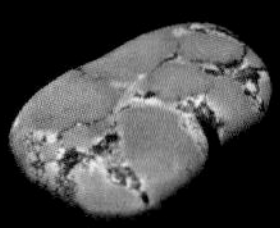

TURQUOISE

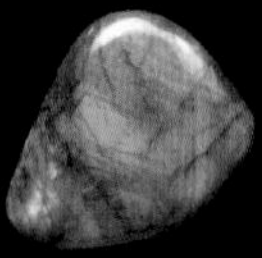

LABRADORITE

APATITE

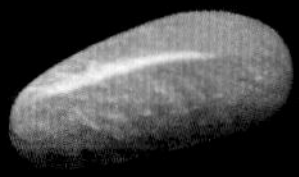

AMAZONITE

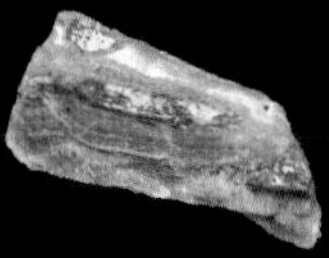

CHRYSOCOLLA

PALE BLUE-GREEN

AQUAMARINE

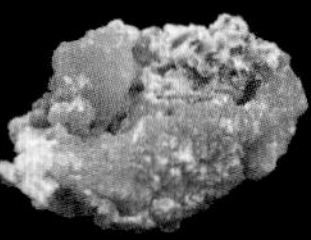

CHRYSOPRASE

PALE BLUE

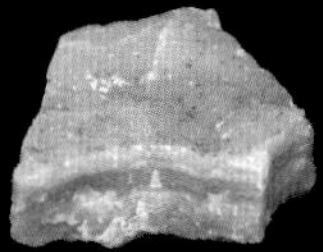

BLUE LACE AGATE

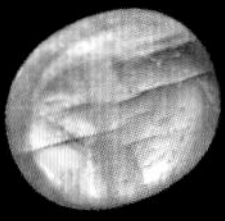

242

BLUE MOONSTONE

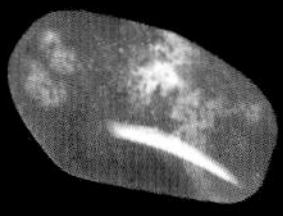

CHALCEDONY

KYANITE

SAPPHIRE BLUE

CELESTITE

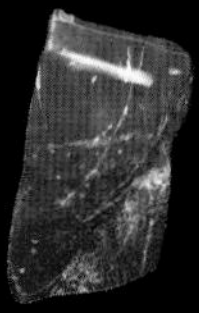

LAPIS LAZULI

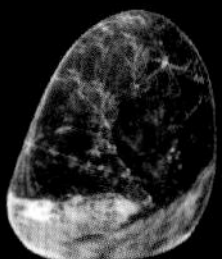

SODALITE

BLUE SAPPHIRE

AZURITE

DARK BLUE

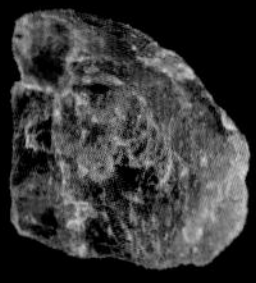

IOLITE

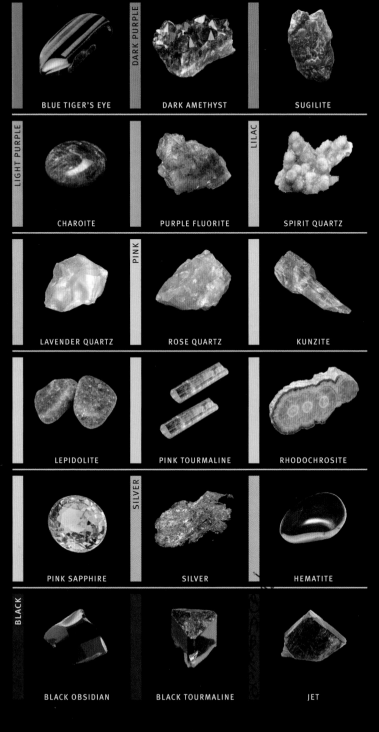

DARK PURPLE

BLUE TIGER'S EYE

DARK AMETHYST

SUGILITE

LIGHT PURPLE

CHAROITE

PURPLE FLUORITE

LILAC

SPIRIT QUARTZ

PINK

LAVENDER QUARTZ

ROSE QUARTZ

KUNZITE

LEPIDOLITE

PINK TOURMALINE

RHODOCHROSITE

SILVER

PINK SAPPHIRE

SILVER

HEMATITE

BLACK

BLACK OBSIDIAN

BLACK TOURMALINE

JET

ONYX

WHITE MOONSTONE

SELENITE

DIAMOND

DANBURITE

AZEZTULITE

PHENACITE

CLEAR QUARTZ

DOUBLE-TERMINATED

244

RUTILATED
CLEAR QUARTZ

ELESTIAL

PHANTOM

LASER WAND

MANIFESTATION

SELF-HEALED

GENERATOR

SOUL-MATE

CHANNELING

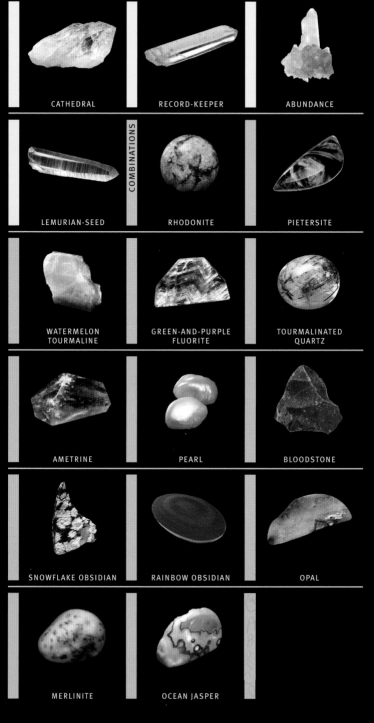

CATHEDRAL

RECORD-KEEPER

ABUNDANCE

LEMURIAN-SEED

RHODONITE

PIETERSITE

WATERMELON
TOURMALINE

GREEN-AND-PURPLE
FLUORITE

TOURMALINATED
QUARTZ

AMETRINE

PEARL

BLOODSTONE

SNOWFLAKE OBSIDIAN

RAINBOW OBSIDIAN

OPAL

MERLINITE

OCEAN JASPER

Color Combinations in Crystals

So far, the crystals covered in this book have had only one color, linking them to individual bands in the color spectrum. In this section, crystals which have different color combinations will be considered. These are special examples, where from a color-healing perspective, both shades contained in the crstyal are significant and have a combined effect. Also, there are examples where more than one colored mineral is present, combining the properties of both to create a differently energized healing tool.

These examples demonstrate the incredible variety of minerals produced by the earth, which can be made

available for healing purposes. Choosing one of these specimens to work with highlights what levels of energy need to be balanced simultaneously; sometimes there is a need to recharge on more than one frequency at a time. Color combinations are a useful visual key to the properties and uses of these stones. However, a person may also be attracted to a combination crystal intuitively, meaning that this is the most appropriate tool at the time. These stones may be worn as jewelry, used in healing, or placed in the home to bring their combined effects into the personal environment of their keeper.

RHODONITE

Rhodonite is an attractive combination mineral with a complicated chemical structure: manganese iron magnesium calcium silicate. Its name comes from the Greek word for rose, *rhodon*. Rhodonite is streaked with black inclusions of manganese oxide; some pieces contain a higher proportion of darker inclusions, and others contain far more pink, allowing a wide variety of selection. It usually occurs with other minerals, including pyrite, calcite, or spessartine. Although rhodonite can occasionally be found in defined crystalline examples, it is much more common in granular massive formations, easily carved into blocks. When polished it has an attractive sheen which resembles deep pink marble.

Containing pink, rhodonite works on the level of the heart chakra, particularly relating to the energy of unconditional love. This is love without sentiment or agendas: it is simple, open and free. This kind of love dissolves all resentment, fear, and anger. The darker areas of rhodonite are symbolic of the earth, meaning that the stone helps neutralize negativity into the vibration of the planet, transforming it through unconditional love. In crystal healing, rhodonite is an excellent tool for strengthening the emotions, helping to support the heart, particularly in relationships where clear communication is vital. It improves the quality of interactions with other people and helps to enhance self-esteem.

RAW

RHODONITE

FORM AND STRUCTURE
triclinic system, rarely found as a crystalline form, more commonly as granular massive formations, with dark inclusions of manganese oxide

COLOR
pink with darker streaks of brown or black

GEOGRAPHICAL SOURCES
Australia, Brazil, Mexico, Russia, Sweden, USA

RARITY
easily obtained

HARDNESS
5.5–6.5

PHYSICAL/EMOTIONAL USES
a gentle but powerful healer of the physical body, supporting liver function and assisting detoxification

soothes the body where there is inflammation, so it is reputed to help rheumatoid arthritis

revitalizes the reproductive system and promotes fertility

heals wounds, whether physical or emotional, and helps to ease shock, anxiety or mental overload

HEALING EFFECTS
place over the heart chakra to ease emotional pain or over any area of the body that is experiencing inflammation

can also be placed over the throat chakra to facilitate the expression of feelings from the heart chakra on the level of unconditional love

PERSONAL USES
a gem remedy made with rhodonite (*see pages 284–85*) eases the symptoms of shock or anxiety

place a piece beside the bed to soothe the mind and improve sleep quality

wear a piece over the heart to enable loving communication

PIETERSITE

Pietersite is a member of the quartz group. It is a mineral that has only recently been discovered, and it has an unusual appearance, made up of swirling masses of mottled layers of dark blue and gold. These layers are combinations of silt, sand, and quartz that have become fused together. The layers also contain fibers that catch the light, but, unlike tiger's eye, where they are laid in parallel bands, in pietersite the bands are folded and mixed, creating a wonderfully irregular play of light through the stone. Pietersite changes color constantly as it is turned toward light, and it also has a luminous, almost metallic sheen.

The dark blue and gold of pietersite are beautiful shades that link the third-eye and solar plexus chakras. The solar plexus chakra holds golden solar energy, light, and vitality; the third eye expands consciousness into wider realms. Looking at pietersite is like observing the contrast between the energy of the sun and the soft depths of the twilight sky. The stone brings brightness, expansion, and flow into new realms of understanding. It encourages the powerful expansion of dreams, ideas, and creative impulses, as well as generating the energy needed to make them happen. It helps to free up energy that is stuck in repetitive patterns, allowing new solutions to emerge.

RAW

PIETERSITE

FORM AND STRUCTURE
quartz in a trigonal crystal system found in masses, with light-reflecting layers

COLOR
dark blue and gold with small white inclusions

GEOGRAPHICAL SOURCES
China, Namibia, South Africa

RARITY
not common; obtain from specialist crystal suppliers

HARDNESS
7

PHYSICAL/EMOTIONAL USES
activates energy, almost in a similar way to a vortex

helps dissolve emotional, mental, or physical blockages, or feelings of being stuck

strengthens the nervous system, and energizes and revitalizes the solar plexus chakra

relaxes muscular tension and gives the body a sense of vitality and renewal

can facilitate inner journeying, allowing visions of other levels of existence with information relevant to the present moment, as well as increasing intuition and clairvoyance

HEALING EFFECTS
place on the solar plexus chakra to revitalize the physical energy of the body

place it over the third eye to activate psychic awareness

works well with gold or blue tiger's eye to enhance its energy

PERSONAL USES
meditate with it to move through blocks and release stagnant energy, whether in the mind, the emotions, or the body

carry a piece to increase creativity and intuitive leaps of awareness

WATERMELON TOURMALINE

The tourmalines are a large group of related minerals, available in a huge variety of colors. The different types have slightly different chemistry, but the most common, called elbaite, is sodium lithium aluminum borosilicate hydroxide. Watermelon tourmaline belongs in this group. It is so named because of its beautifully distinctive combination of a green outer band with a pink center, which looks very similar to the colors inside a watermelon. Like all tourmalines, this mineral is pleochroic, meaning that it looks deeper in color when viewed from its narrow end rather than its side. It is also piezoelectric, meaning that squeezing or vibrating a crystal will create a different electrical charge at each end.

The combination of vivid green and pink in watermelon tourmaline is the most effective for balancing and healing any issues in the heart chakra. Green is the color first associated with that chakra, the color of growth and expansion. Pink is the color of unconditional love, the higher frequency of energy also associated with the heart chakra. Watermelon tourmaline expands and releases the heart into the fullness of its potential, enabling a joyful expansion into all that life has to offer, especially in terms of relationships with others. It enhances the senses, triggering a fresh set of feelings with which to interpret the world. The lithium content in watermelon tourmaline makes it helpful to the mind, encouraging relaxation and relief from stress.

RAW

WATERMELON TOURMALINE

FORM AND STRUCTURE
trigonal crystal system, forming long prismatic crystals with striations running along the vertical axis, often found embedded in granite

COLOR
pink center surrounded by green bands; tourmalines also occur in a huge variety of colors: black, brown, green, pink, yellow, red, or violet

GEOGRAPHICAL SOURCES
Brazil

RARITY
obtain from specialist crystal suppliers

HARDNESS
7–7.5

PHYSICAL/EMOTIONAL USES
eases and soothes all issues of the heart, whether linked to the physical organ or to the chakra center

calms emotional stress, fear, or anger and brings relaxation

opens the heart to love, whether in relationships or on the wider unconditional level, where it extends to nature and to the planet as a whole

stimulates an appreciation of the beauty in nature

HEALING EFFECTS
place over the heart chakra to balance the energy there, and help it to open more fully

PERSONAL USES
wear over the heart chakra to soothe and balance the emotions and to help bring love into your life

GREEN-AND-PURPLE FLUORITE

Fluorite is one of the most common minerals, available in a great variety of different colors. It has a cubic structure, forming well-defined crystal shapes and is composed of calcium fluoride. This mineral gives its name to the light effect called fluorescence, because fluorite crystals glow when exposed to ultraviolet (UV) rays. Green-and-purple fluorite is a mixture of two of the most common colors represented by this crystal group. Every piece is individual, some with bands of the two colors, others with a flowing interplay of the different shades. Polished green-and-purple fluorite is beautifully translucent with a soft luster; it is especially attractive when set in jewelry.

The complementary shades of purple and green in this type of fluorite are linked to the heart (green) and the crown (purple) chakras. Green is the color of expansion, growth, and reaching toward the light. Purple is the shade associated with spiritual expansion and awareness of higher levels of consciousness. The two combined enable the wisdom of spiritual awakening to come into the physical body and reside in the heart. Meditating with green-and-purple fluorite brings a gentle flowing sense of peace to the whole system—body, mind, and spirit. This is the ideal state to be in to experience spiritual expansion and unfolding into new levels of awareness. Peace of mind and heart is also deeply relaxing to the entire self, enabling the nervous system to recover from the strenuous demands of everyday life.

POLISHED

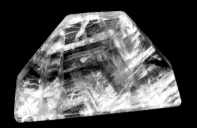

GREEN-AND-PURPLE FLUORITE

FORM AND STRUCTURE
isometric system, forming well-shaped cubic crystals, often in octahedrons

COLOR
purple and green, also in individual shades of green, purple, pink, yellow, red, white, and black

GEOGRAPHICAL SOURCES
Brazil, Britain, Canada, China, Germany, Mexico, USA

RARITY
easily obtained

HARDNESS
4

PHYSICAL/EMOTIONAL USES
helps to calm and relax body and mind, and ease physical or emotional stress

balances the mind in line with the emotions enabling clear thinking and decision making

cleanses and replenishes the energy field around the body

HEALING EFFECTS
place over the heart chakra to ease emotional stress, or over the third-eye or crown chakras to help bring replenishing energy to the whole system

place a piece in each hand to balance and restore the whole energy field

PERSONAL USES
wear to bring peace and joy into your life and enhance relaxation

TOURMALINATED QUARTZ

This beautiful and unusual mineral is a combination of quartz with inclusions of black tourmaline, which is also called schorl (sodium iron aluminum borosilicate hydroxide). The best examples come from Brazil; they vary from clear quartz to snow quartz, with the black tourmaline in strands or visible lumps. The quartz is the dominant crystal form, in its typical hexagonal system; however, the different linear structure of the black tourmaline is visible within it.

This melding of these two minerals creates a powerful combination in healing terms. Clear quartz is an amplifier, one of the most common and powerful crystal-healing tools, used to increase energy in the system. Black tourmaline acts like a conductor, picking up any stray elements of negativity and sending them into the neutralizing zone of the earth. This combination makes tourmalinated quartz one of the most powerful cleansing and purifying crystals to use in a healing context. It also exerts a powerful effect on an environment, helping to neutralize high levels of radiation or electromagnetic energy that may be causing stress to the body or its systems. Keep it by a computer or on top of a television, for example, to neutralize unwanted electromagnetic waves.

Tourmalinated quartz also aligns the chakra system, and integrates it into the earth's own grid, helping to alleviate symptoms of environmental stress that can be caused by constant bombardment from radio waves, microwaves, and cell phone masts.

RAW

TOURMALINATED QUARTZ

FORM AND STRUCTURE
trigonal system with inclusions of schorl (black tourmaline) in visible strands or lumps

COLOR
clear or white with black stripes

GEOGRAPHICAL SOURCES
Brazil

RARITY
obtain from specialist crystal suppliers

HARDNESS
7

PHYSICAL/EMOTIONAL USES
an exceptionally powerful purifying and cleansing crystal, eliminating physical and psychic toxins from the system

helps to neutralize effects of electromagnetic stress

restores the balance of the entire chakra system and realigns the energy field with the natural geopathic grid of the earth

clears the mind of clutter or negativity and establishes clear patterns of thought

clears negative emotional issues relating to the past, enabling a fresh start to be made

HEALING EFFECTS
in healing layouts, place over the top of the head and between the feet to enable a flow of energy throughout the whole system

PERSONAL USES
place in the home, by a computer or television, or in a healing space to cleanse and purify the environment

AMETRINE

This unusual mineral is a mixture of two types of quartz in the same stone—purple amethyst and yellow citrine. Both crystals are types of quartz colored by different amounts of iron. If amethyst is heated, it turns into yellow citrine, so the variation of color in ametrine suggests exposure to different levels of heat during the natural formation process, meaning both colors appear in the same crystal. A single piece of ametrine shows bands of both amethyst and citrine swirling and melding together, creating a beautiful soft play of colors. This is particularly evident when ametrine is polished.

Amethyst is associated with the crown chakra and the spiritual realm, especially in relation to personal guides, angels, and sources of expanded awareness. Citrine's golden yellow color connects it to the solar plexus, to the center of personal will. The meaning of ametrine is the alignment of the human being with the spiritual "higher self," bringing both into harmony. In practical terms, this means listening to and following intuitive feelings, encouraging creativity and new directions, and applying the practical mind to make things happen. Ametrine aligns the human mind with spiritual consciousness, which is expressed in the phrase: "Not my will but thy will be done." It allows energy to move through the chakra system in order to encourage full expression of cosmic energy on earth.

POLISHED

AMETRINE

FORM AND STRUCTURE
trigonal crystal system, forming quartz crystals comprised of both amethyst and citrine

COLOR
purple and yellow-gold

GEOGRAPHICAL SOURCES
Bolivia, Uruguay

RARITY
obtain from specialist crystal suppliers

HARDNESS
7

PHYSICAL/EMOTIONAL USES
helps encourage the balance between intuition and the practical mind

increases creativity and improves concentration

helps keep the mind in focus during study or intense work pressure and improves the ability to solve problems

improves metabolism and digestion of ideas as well as food

helps to regulate body weight

is reputed to enhance the immune system and support people experiencing long-term chronic postviral symptoms

HEALING EFFECTS
place pieces over the crown and solar-plexus chakras to bring these two centers into balance with each other

adding a piece of rose quartz over the heart chakra allows that center to be in deeper harmony with the mind

PERSONAL USES
wear or carry ametrine to stimulate creativity on all levels and also the concentration to bring ideas into practical use

use to bring spiritual awareness into everyday life

PEARL

Iridescent pearls are organic in origin, forming inside the shells of a sea creature called the pearl oyster. The intrusion of some kind of foreign body irritates the inner lining of the shell, and the creature secretes layers of calcium carbonate, as well as a special secretion called conchiolin. The combination of these two ingredients creates layers of nacre, the proper name of mother-of-pearl. It can take several years to produce a single pearl within the shell.

Pearls are either freshwater, produced mostly in China, or saltwater, produced in the Persian Gulf or Indian Ocean coast of Australia, as well as Japan, Fiji, Tahiti, and Indonesia. "Produced" is a key word here because the vast majority of pearls these days are cultured, meaning they are created by deliberately introducing a bead into the mollusk to stimulate the secretion of the pearl-forming nacre. Cultured pearls tend to be spherical, whereas natural pearls are slightly irregular in shape and extremely rare. The only way to confirm if a pearl is natural is to X-ray it, which will show the presence of an artificial bead under the layers of nacre if it is not natural.

In healing terms, pearls have a gentle energy, symbolic of the moon and its influence on inner cycles within the body.

CULTURED PEARLS

PEARL

FORM AND STRUCTURE
organic mineral with an amorphous structure, formed within the shell of the pearl oyster caused by irritation or the introduction of a bead to create a cultured pearl; layers of nacre (calcium carbonate plus natural secretions) creates layers of iridescence with rainbow colors

COLOR
white-cream, yellow, green, blue-gray, purple, brown, black

GEOGRAPHICAL SOURCES
Australia, Fiji, Indonesia, Japan, Persian Gulf, Tahiti (saltwater), China (freshwater)

RARITY
rare as mineral samples; mostly used in jewelry

HARDNESS
2–2.5 (soft, scratches very easily, keep wrapped in soft cloth)

PHYSICAL/EMOTIONAL USES
pearl's energy helps to improve hormone balance and bring the feminine cycle in line with the moon, making it helpful for fertility purposes

harmonizes the emotions and eases mood swings

balances water within the body, easing fluid retention

brings emotional peace and spiritual tranquility

HEALING EFFECTS
place over the sacral chakra (influencing the reproductive organs), the throat chakra (influencing the thyroid), and the brow (influencing the pineal and pituitary glands, which control the hormones) to balance hormonal activity

PERSONAL USES
wear as jewelry to experience peace and tranquility

BLOODSTONE

Bloodstone is a variety of chalcedony quartz with a distinctive dark green color combined with specks and clumps of red jasper inclusions. Being a member of the quartz family gives it a hardness suitable for carving, and it is often used as an ornamental stone for inlays alongside other types of quartz or marble. The distinct blood red marks within the stone have made it a traditional talisman, believed to protect against evil spirits. In medieval times, the specks were said to be the blood of Christ that had dropped down to the ground when he was crucified.

In crystal healing, this stone is often linked to male energy; sometimes it is referred to as a warrior stone because its red specks resemble blood. Red is the color of the base chakra, symbolizing action, physical strength, and vitality, the power to make things happen in the world. The basic green color of bloodstone links it to the heart chakra and expansion into love. Legends such as the tale of Percival, one of King Arthur's knights who went on the quest for the Holy Grail, illustrate the combined qualities of bloodstone. Percival was a knight trained in the physical aspects of combat, but his quest took him into the spiritual realms, teaching him lessons about illusion and the power of unconditional love. Meditation with bloodstone helps before embarking on any new venture, as it brings the energy of the heart into balance with the energy of action.

POLISHED

BLOODSTONE

FORM AND STRUCTURE
trigonal system found in massive formations, with inclusions of red jasper

COLOR
green with dark red or orange-red specks or stripes

GEOGRAPHICAL SOURCES
Australia, Brazil, China, India, USA

RARITY
easily obtained

HARDNESS
7

PHYSICAL/EMOTIONAL USES
strengthens the body, cleanses the blood, and brings vitality to the system

helps maintain stamina during periods of intense physical activity and gives the strength to overcome obstacles

encourages belief in oneself and commitment to spiritual goals

promotes good physical and emotional health

HEALING EFFECTS
in healing layouts, place over the heart, with a piece of clear quartz over the brow and a piece of smoky quartz between the feet— this provides a combination of crystals in order to stay centered and receive a vision of the pathway that opens up to new life experiences

PERSONAL USES
wear or carry when courage or stamina in pursuit of personal goals is needed

SNOWFLAKE OBSIDIAN

Obsidian is a form of volcanic glass that cools very quickly when lava flows come into contact with water. This produces a glassy-textured mineral with a tendency to break into flakes when struck in a particular direction. This gives a razor-sharp edge, so obsidian tools, blades, and arrowheads have been found dating back thousands of years. Obsidian is mainly composed of silicon dioxide (quartz), and it contains many impurities, including iron and magnesium, giving it a dark color. Snowflake obsidian is black with white inclusions made up of other minerals such as mica or feldspar. These look like tiny snowflakes, which is the reason behind its descriptive name.

The mineral's black background symbolizes the absorbing power of darkness, the place of mystery and magic. The white inclusions are like the speck of white in the black half of the yin-yang symbol, which is there as a reminder that within darkness there is always light; the other half of the symbol shows that within light there is always dark. This is an eternal truth, that one is always changing into the other and back again. Snowflake obsidian illustrates the fact that both sides of the coin, both black and white, have their lessons, and both are the most extreme polarities in the spectrum. They belong side-by-side, opposite but equal.

POLISHED

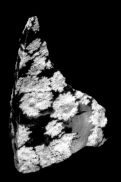

SNOWFLAKE OBSIDIAN

FORM AND STRUCTURE
volcanic glass rich in silica, stained by impurities with inclusions of mica or feldspar minerals making whitish marks

COLOR
black with white specks

GEOGRAPHICAL SOURCES
Britain, Mexico, USA

RARITY
easily obtained

HARDNESS
5–5.5

PHYSICAL/EMOTIONAL USES
transmutes negativity and provides energy for creative problem solving

protects the root and sacral chakras, helping to preserve life force and vitality in the body as well as strengthening the aura

acts as a shield against geopathic or environmental stress

shows the light at the end of the tunnel after periods of challenging experiences, so restoring a positive outlook

HEALING EFFECTS
place over the root chakra to help ground and stabilize its energy

place a piece on each side of the body to act as an energy shield

add red tiger's eye to the root chakra as well to warm and stimulate the area

PERSONAL USES
wear or carry to protect the aura from environmental stress

meditate with snowflake obsidian to receive creative energy for problem solving

RAINBOW OBSIDIAN

All forms of obsidian are made up of silica-rich volcanic glass that cooled rapidly in the presence of water. The result is an unstructured mineral that breaks into sharp-edged pieces. Many thousands of years ago, humankind used pieces of obsidian for making cutting tools. Rainbow obsidian is a rare form of this volcanic glass with a very distinctive quality. When a piece is held toward a strong light source, bands of metallic-looking rainbow colors appear in shades of pink, gold, blue, and green. This special effect occurs because of tiny specks of crystalline impurities and trapped microscopic air or water bubbles within the stone, all of which reflect light. The end result can be really dramatic, especially if pieces are carefully cut and polished. It creates the impression of a rainbow pool of colors.

Rainbow obsidian is used in crystal healing as a symbol of all the steps in transformation that are needed for a human being to become a vessel for receiving light and unconditional love. The colors in rainbow obsidian swirl out of darkness, symbolizing the infinite magical and mysterious potential of the unknown. The colors shift and change, stimulating different levels of awareness. This mineral causes colorful and evocative dreams that may solve personal issues or questions.

POLISHED

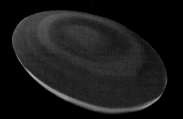

RAINBOW OBSIDIAN

FORM AND STRUCTURE
amorphous volcanic glass, rich in silica with many impurities; water and air bubbles also contribute to the light reflections

COLOR
black until turned to a light source, then becomes filled with pink, gold, blue, and green iridescent bands

GEOGRAPHICAL SOURCES
USA

RARITY
obtain from specialist crystal suppliers

HARDNESS
5–5.5

PHYSICAL/EMOTIONAL USES
a powerful activator of the unconscious mind, stimulating vivid and colorful dreams with symbolic messages

assists recovery from physical or emotional trauma, ensuring that strong emotions, such as fear, do not get trapped inside the physical body

helps to bring issues with close family or previous generations to the surface so that these can be cleared

is reputed to help the healing of broken bones

HEALING EFFECTS
place over the third-eye chakra to help bring deep emotional or family-related issues to the surface for healing; add a piece of smoky quartz between the feet to ground the body and a piece of rose quartz over the heart to ease painful feelings

PERSONAL USES
wear or carry to stimulate intuition and creativity in life

OPAL

Opals have been prized gems for centuries because of their magically beautiful play of multicolored layers of light. Opal is classed as a mineraloid, because it does not have a truly crystalline structure.

The chemistry of opal is mainly silicon dioxide, but there is also water present (between 5 and 10 percent). The molecules containing silicon and oxygen atoms are enclosed in extremely tiny spheres, and these capture and refract light, giving opal its flashing appearance of rainbow colors, a phenomenon called opalescence. There are many kinds of opal, ranging from the more familiar pale blue, turquoise, and green shades to other examples such as the cherry opal, found in red, pink, or orange, and the fire opal, which has red-orange sparkles. The white opal, milky with gentle rainbow iridescence, is perhaps the best known. If an opal dries out too much, it can crack or shatter because of water loss, so it needs to be stored carefully.

The word *opal* derives from the Sanskrit word *upala*, meaning precious stone; in India the stone is regarded as having magical properties.

The Romans mined opals in what is now the Czech Republic. Fire opals come from Mexico, and were originally brought to Europe by the Spanish conquistadores in the sixteenth century.

WHITE OPAL

OPAL

FORM AND STRUCTURE
amorphous mineraloid found in massive formations or as cavity fillings in fractures or pockets within bedrock

COLOR
white, blue, pink, red, black, orange-red, yellow, with different levels of rainbow opalescence

GEOGRAPHICAL SOURCES
Australia, Mexico, USA

RARITY
rare as a collectable mineraloid (consult specialist crystal suppliers); often used in jewelry

HARDNESS
5.5–6 (may dry out and become brittle)

PHYSICAL/EMOTIONAL USES
opals are found in many different colors and can be matched to the chakra to which they correspond, for example, red to the root or blue to the throat

generally, in crystal healing, opals are all seen as emotionally supportive, stabilizing mood swings and helping to increase trust in one's feelings

the more fiery-looking opals increase vitality, passion, and zest for life

HEALING EFFECTS
in healing layouts, match pieces of opal to the chakra to which they correspond by color, for example, blue to the throat to open communication

white opal is useful in gently balancing the crown chakra and opening the gates to spiritual expansion

PERSONAL USES
wearing opal is sometimes considered to be unlucky unless you are born in Libra, for whom it is a birthstone; however, this is just superstition—if you are drawn to an opal, then go ahead and wear it

MERLINITE

Merlinite is not a proper mineralogical name; it has been assigned via the crystal-healing domain as a comment on the perceived properties and uses of this stone. Its composition is a mixture of quartz and another mineral called psilomelane (manganese oxide), which forms black layers, streaks, or specks within the main crystal structure. The overall effect is usually a white background streaked with black, although occasionally it is found as borytroidal black psilomelane coated with quartz crystals, which is a much more unusual specimen.

In crystal healing, merlinite is viewed as a powerful stone of magic. Its name alludes to the magician Merlin; thousand-year-old legends tell that he was a teacher to the young Arthur before he became king. Merlin's domain was the deep forest, the place where light and shadow interplayed and where the veils between levels of reality were thin. In legend, Merlin was able to change shape, transforming into animals or birds, and this was a sign of his shamanic power. Shamanism is the oldest form of spirituality on the planet, still practiced today in many cultures, including the Inuit and the medicine men of the Amazon rainforest. It involves a deep relationship with the energy field that surrounds all nature, dissolving the bonds of ordinary three-dimensional reality. Merlinite is a crystal that helps facilitate shamanic development.

SHAPED AND POLISHED

MERLINITE

FORM AND STRUCTURE
quartz masses in a trigonal system with inclusions of psilomelane as layers, specks, or stripes

COLOR
white with black specks, stripes, or bands

GEOGRAPHICAL SOURCES
USA

RARITY
obtain from specialist crystal suppliers

HARDNESS
7

PHYSICAL/EMOTIONAL USES
merlinite is a stone of transformation that works on levels beyond the rational mind, so inviting the energy of magic into everyday life and bringing about radical changes of perception and circumstances

teaches how to integrate darkness and light, a necessary step in spiritual development

helps move through blockages that hold back growth, such as fear of change, fear of new things, and fear of the unknown

HEALING EFFECTS
in healing layouts, if merlinite is placed over the third-eye chakra, it helps open psychic perception in a powerful way

the effects of this stone are best explored with the help of a professional crystal healer

PERSONAL USES
a small piece of merlinite is useful for meditation to open up awareness of other realms in nature

OCEAN JASPER

This is a recently discovered mineral which has several interesting characteristics. It is a new and unusual form of jasper, which is a type of microcrystalline quartz. It has an extremely unusual appearance because it is covered in multicolored circles, specks, and swirls. Ocean jasper is associated with rhyolite, an igneous rock similar to granite. This is why ocean jasper is sometimes referred to as petrified rhyolite; another common name for it is orbicular jasper because of its many circular ripples and shapes.

Ocean jasper's appearance, like a series of radiating circles, is the inspiration for its use in healing, where it is seen as a stone of joy, expansion, and relaxation. When polished it has the typically silky-smooth feel of jasper, making it a wonderful stone to hold. It contains all types of colors, including pink, gold, green, black, white, yellow, and brown, usually in softer more muted shades. These symbolize a gentle and supportive action, working on the unconditional-love level in the heart chakra, as well as stabilizing the solar plexus and sacral chakras. It celebrates the circle, over and over again, the symbol of the never-ending cycles of life.

RAW

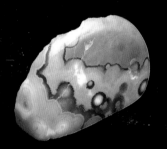

Dark Amethyst

Calms mental stress
Eases headaches
place under pillow to
ease sleep

A

TREATED QUARTZ

This is natural quartz that has been treated in a special way to bring out certain vivid shades of iridescent color. The crystal is coated with a metal delivered in the form of a vapor. This creates a fine layer over the top of the crystal surface that is barely a few atoms thick. Crystal healers like to use treated quartz because of the bonding between precious elemental metals and pure natural-quartz crystals. In crystal-healing layouts, it is useful to place a piece of treated quartz on its corresponding chakra color.

RUBY AURA

Another variety of treated quartz, called Ruby Aura, is very similar. The crystal is coated with a mixture of gold and platinum vapor that turns it a rich red. Ruby Aura energizes the root chakra, adding the expansive solar energy of gold and the brilliant starlike quality of platinum. It brings energy to the physical body and encourages passion and zest for life. It helps to revitalize the body after an illness, particularly when there are feelings of extreme tiredness or weakness.

OPAL, ANGEL, OR RAINBOW AURA

A pale opaline iridescent type of treated quartz is called Opal Aura, Angel Aura, or Rainbow Aura. This is quartz coated with platinum on its own, creating a soft and subtle sheen, a play of rainbow colors, and yet still the transparent crystal shines through. It has a quality of sweetness that lifts the spirits. This variety is said to connect the wearer to the subtle angelic realms, bringing in energy at a very high but gentle frequency. It helps to keep Angel Aura in a meditation space to facilitate connection with a personal guardian angel.

GOLDSTONE

This is a very special kind of treated glass, made using a technique perfected by the Miotti family in Venice in the seventeenth century. Its origins were shrouded in legends and mystery; local folklore linked it even further back to medieval times and an obscure monastic order, giving it the name monk's gold. Its other name *stellaria* refers to its starlike reflections. In Italian, another common name for this stone was *avventurina*; this name became the term used to describe the mineral aventurine, with its sparkling mica inclusions (*see pages 104–5*). Some books refer to goldstone as aventurine glass, but this name is now confusing, so goldstone is the preferred modern term.

During the glassmaking process, pure copper is melded into the matrix to produce a swathe of microscopic sparkling spheres. Some goldstone is an orange-yellow color, and other types have a dark blue or violet background. The manufacturing process of goldstone was a closely guarded Miotti family secret for hundreds of years before being brought to light in the nineteenth century by another Italian, Pietro Bigaglia, a maker of filigree glass. Goldstone is even more beautiful when smoothed and polished; it is used to make beads, jewelry, or even small figurines.

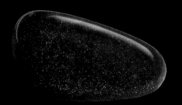

GOLDSTONE

STRUCTURE

Some crystal healers reject the idea of using goldstone as a healing tool because it is, essentially, glass. Although it contains silica, it has no internal lattice structure to make it crystalline, so therefore it may not be effective for working with energy. Its glass background is pure silica, and it has an amorphous structure. It does, however, contain microscopic copper spheres, and therefore can be seen as a useful way of bringing the conductive energy of copper into contact with the energetic field of the body as a sparkling shower of particles.

QUALITY

The most attractive quality of goldstone is its glittering sheen, and perhaps one of its ancient names, *stellaria*, holds a key to its use. Particularly in its dark blue form, goldstone looks like the deepest blue night sky patterned with stars. A polished piece has a silky feel and a deep luster; meditating with it is like going on a journey into the depths of space because all its microscopic particles sparkle like galaxies and constellations. Placed over the third eye between the eyebrows, it enhances inner journeying, bringing visions of other worlds.

Crystals
in Your Life

Crystals add wonderful qualities to life: their beauty enhances personal space, their colors attract the eye, their energy communicates a sense of presence. In this section, ways to choose and care for crystals will be shown, as well as many different approaches to their practical use. Building a collection of crystals is a very satisfying hobby; understanding their significance as well as their physical attributes adds new levels of fascination to the subject.

By placing crystals in your home or work environment, or by carrying them or wearing them as jewelry, we can find many more ways of experiencing their helpful effects.

Meditation with crystals and making crystal remedies are further examples of ways in which crystals can be useful, sensitive tools.

Simple approaches to crystal healing will also be shown as a starting point. These are intended for use with a friend as a way of exploring how crystals affect energy. This requires sensitivity and openness, as well as trust and helpful communication between both parties, so that the experience of using the crystals is positive and beneficial. Expressing what feels good and what does not is an important part of the exchange. Working with crystals encourages the development of new levels of intuition and creativity

Choosing, cleansing, and caring for crystals

Crystals can be bought in many different locations, including New Age stores, specialist crystal suppliers, and crystal exhibitions or fairs. The more specialized the outlet, the bigger the choice of crystals available and the likelihood that stock will be of good quality. Serious crystal collectors will look for a reputable supplier who knows the geographical origin of the stones as well as correctly identifying them.

Different types of crystal are typically available:

CRYSTAL CLUSTERS • these are made up of many small crystals emerging from a common base; a good example is amethyst

SINGLE CRYSTAL POINTS • these tend to be pieces of quartz, and a good supplier will name them individually if they have specialist labels such as phantom, cathedral, generator, etc. (*see pages 230–37*)

LARGE UNPOLISHED PIECES • common examples are rose quartz or aventurine, and these work well placed outside

TUMBLESTONES • these are small shiny crystal pebbles polished with gravel; they are the cheapest form of crystal and are popular with children

SHAPED CRYSTALS • these can be pieces of quartz that have been carved into particular shapes, such as the rose quartz or citrine flame shaped pieces that are becoming popular, or carved polished spheres in different sizes

MINERALOGICAL SPECIMENS • these are often delicate—sulfur or copper, for example—and may be presented in boxes

For healing purposes, small clusters, single points, or tumblestones are the best choice.

SINGLE CRYSTAL POINT

UNPOLISHED

TUMBLESTONES

CLEANSING CRYSTALS

When you buy a crystal it is a good idea to wash any dust off it with lukewarm soapy water and then pat it dry. This simply cleans away surface dust, but there is another level on which crystals need to be cleansed using energy, which prepares them for use in healing and also programs them to your individual frequency. There are two steps to this. First, use one of the following methods to clear the crystal matrix:

• place the crystal under cold running water for a few moments
• strike a tuning fork or small bell beside it to cleanse it with sound
• light some incense and pass it through the smoke.

Second, hold the crystal in your left hand and place your right hand over the top of it. Then close your eyes, breathe deeply, and focus your energy. Say: "May the energy of Divine Love infuse this crystal so it works for the highest good."

This process clears and prepares the crystal to amplify and enhance energy beneficially in your life.

283

CARING FOR CRYSTALS

Crystals need to be looked after, so when you buy a stone, check its profile. Here are some simple tips to bear in mind:

• if it is soft, like amber, it needs to be kept away from harder crystals that could scratch it

• avoid knocking quartz crystal points or they will shatter

• many crystals, such as amethyst, will fade in color if exposed to strong sunlight, so they are best kept in the shade

• small crystals—Herkimer diamond or spinel, for example—are best kept safe in a box

• precious stones, such as opal, and organic minerals, such as pearl, need to be kept wrapped in a soft cloth and away from heat to prevent scratching or shattering.

AMETHYST

OPAL

AMBER

Making crystal remedies

Crystal remedies are a subtle way of experiencing the properties of crystals in water. The method of making them is similar to the one used by Richard Bach, the originator of the Bach Flower Remedies. He put petals and leaves in spring water and left them in the sun, so that the sunlight would enable the transfer of the energy signature of the plant into the water along with its healing properties. The same process can be used with many crystals, especially those in the vast quartz family. A crystal remedy can be taken to support the system in addition to using physical methods with crystals, such as healing layouts.

There are two types of preparation. One is energized water, and the other is a crystal remedy.

Energized water • This is made simply by placing a cleansed crystal, such as a piece of clear quartz or amethyst, into a clear glass jug of spring water and leaving it in direct sunlight for an hour. Then take the crystal out and drink the water throughout the day to receive the benefits.

BELOW *Drinking energized water is a lovely way to experience the healing properties of crystals. It helps to rebalance your energies and calm your mind.*

A remedy • This is made by placing a cleansed crystal in a glass bowl of spring water and leaving it in direct sunlight for three hours. Take the crystal out and pour the energized water into a large dark-glass bottle so that it is half-full of energized water, and top it up with brandy.

This is a mother remedy. Shake the bottle and place it in a dark cupboard for two weeks, shaking periodically to maintain the energized potential. To make the final remedy, fill a small dark-glass bottle up to two-thirds with spring water and top up with the mother remedy. Shake this bottle and you have the stock bottle you take as a remedy: two or three drops in a glass of water three times daily. A bottle of mother remedy can be kept indefinitely and produces many stock bottles.

As well as drinking remedies in water, they can be added to the bath to enhance relaxation—use four drops per bath. They can also be dabbed on the chakra point corresponding to the crystal's color to energize that area. Another suggestion is to make an energizing and cleansing room spray in a pump-action spray bottle. Add two-thirds spring water and one-third remedy, shake and use.

CRYSTAL REMEDIES

Some common crystals to use for making remedies are:

ROSE QUARTZ • gently loving and supportive, eases emotional problems, good for children

AVENTURINE • soothes the heart and enhances relaxation

CITRINE • energizes the system and enhances mental clarity

SMOKY QUARTZ • cleanses away negativity; can also be used in a spray bottle to cleanse the environment

AMETHYST • enhances spiritual awareness and relieves mental stress

BLUE LACE AGATE • softly relaxing, soothing; brings peaceful dreams

Crystal meditation

Meditation is an art that balances the entire chakra system, opening the way to inner peace and harmony. There are many different forms of meditation, all of them offering the potential to reach a state of spiritual enlightenment. Meditation in its purest sense is not allied to any particular religion; it is simply a practice that enables spiritual development. It is found in the teachings of many ancient traditions, including Zen Buddhism and Hinduism.

Meditation is not a quick fix, it is an art that takes time to learn and absorb. Practiced over a period of time it brings real benefits, such as lowered blood pressure, a calmer mind, and less turbulent emotions. Regular practice is the key to success. Twenty minutes a day is a good starting point, and the best time to do it is first thing in the morning. Either sit on a hard-backed chair with your legs uncrossed and feet on the floor, hands loosely clasped in your lap, or sit cross-legged on the floor in a meditation posture.

FOCUSING ON A CRYSTAL

Using a crystal is another aspect of meditation practice where the energy of the crystal itself can contribute to the experience. Selecting a crystal for meditation is entirely your own choice. You may be drawn to a stone because of its color, shape, or feel. You may have read about a stone and decided you want to work with it. You may simply look at your collection and feel that one particular piece jumps out at you.

There are two simple ways to meditate with a stone. The first is to gaze at it. Meditation with eyes focusing on an object is a technique from Zen tradition, where contemplation stills the mind. If you decide to try this method, set the stone on a plain cloth on a clean surface, with a lit candle beside it. Settle yourself in the meditation pose and breathe rhythmically as you relax. Let your eyes rest on the stone and focus on it for as long as you can without looking away. If you need to do so, look at the candle for a moment, then back at the stone. Notice any thoughts, feelings, or impressions that come to you and then let them flow away. After your twenty minutes, write notes to recall any important details.

MAINTAINING BALANCE

The second way is to hold the crystal in your left hand and cover it with your right. The left hand is receptive and the right hand dynamic, and by holding the crystal between them in this way the yin-yang balance of energy is maintained. Relax, close your eyes, and breathe calmly. Simply let your mind rest and focus on the stone between your hands. Notice how it feels, how you feel. You may notice sensations of warmth or tingling in your hands or elsewhere in the body. This is the energy of the crystal interacting with your own. After breathing deeply and coming out of the meditation, make notes of any important impressions.

ABOVE *Holding crystals is a way to relate to them individually; you may feel warmth or tingling in your body, in chakra areas such as the third eye or the crown.*

Crystals in the environment

Crystals can play a big part in maintaining positive clear energy in the environment. They can be placed indoors or outdoors in order to enjoy their effects. Deciding where to put a crystal is an individual choice, and intuition will guide you to the most appropriate location. Some crystal healers say stones like to be moved around; every so often you may sense a need to vary their position—and if so, just do it!

INSIDE

Inside buildings, crystals bring sparkling light to a space, as well as different colors and also the special properties they hold. Here are some ideas to try in different rooms:

In the office/workspace • Computers, telephones, or mobile phones that receive and emit electromagnetic and radio waves are sources of environmental stress. Placing neutralizing and cleansing crystals such as smoky quartz, black tourmaline, jet or obsidian in the office, on top of computers or phones, will absorb negative waves or radiation and channel them back into the earth.

JET

RAINBOW FLUORITE

In a bedroom • Keeping crystals in the bedroom is a matter of personal choice. Some, including zincite, have a lively energy and might interfere with sleep patterns. Even clear quartz crystals can be too intense for some. Softer versions of quartz, such as amethyst, rose quartz, or aventurine, have a gentler effect and can improve your sleep if placed under the pillow.

ORANGE CALCITE

BLUE CHALCEDONY

In a meditation space • If you have a space dedicated to meditation, yoga, or healing, then placing crystals is a wonderful way to enhance the energy there— again, use your intuition to guide your choice. For example, clusters of quartz or amethyst, pieces of orange or green calcite, blue chalcedony, or colorful green-and-purple fluorite will all share their healing energies in the space, preparing for relaxation and spiritual development.

OUTSIDE

Outside in the open air, crystals can also add special energy. Here are some ideas for placing crystals in outside spaces:

In the garden • Crystals can be placed among plants to enhance their energies—green stones, such as aventurine or moss agate, are particularly good to use because of their color and their reputed ability to enhance the energy matrix of nature. Large pieces can be obtained, and these are particularly suited to outdoor spaces where they become special features in the garden landscape.

In water • Water features are becoming more popular in gardens, and adding combinations of crystals to a fountain or flowing display adds even more color and sparkle. Rose quartz, green or orange calcite, or a display of multicolored pieces of jasper is lovely to try. You can also tend your plants with the energized water!

In a sacred space • Special garden spaces are also becoming popular, such as circular paved areas, maybe with a fire pit, or a planted area around a hot tub, or a special outdoor meditation space. Small crystal tumblestones lend themselves to the creation of outdoor mosaics in order to completely individualize a space.

ABOVE *Sparkling with light, crystals bring an enhanced sense of beauty to an outdoor space, and create an atmosphere of peace and harmony.*

Crystals for children

Children are naturally fascinated by crystals because of their colors, shapes, and textures. Encouraging exploration of the world of crystals helps children to learn about the earth and its dynamic nature, the way crystals are formed, and the amazing diversity of minerals that are found in the earth's crust. Starting children off with a small collection and teaching them the name and significance of each stone can ignite a lifelong passion. It can also encourage them to become more interested in the natural world around them, so that even simple stones on a beach become worth collecting as beautiful objects.

Semiprecious stones are widely available. Children often respond to stones according to their favorite colors and may like to collect different examples of one color shade—in the pink range, for example, rose quartz, pink calcite, lepidolite, and rhodonite.

Showing children how to clean, care for and keep stones early on in life leads to a satisfying hobby. A bag or box designated to keeping stones in reinforces the idea that crystals are precious, special and worthy of respect.

BELOW *Children are naturally drawn to the colors and textures of crystals; starting a collection can stimulate a lifelong interest.*

A CHILD'S COLLECTION

Here is a simple kit of ten inexpensive stones suitable for starting a small collection:

RED JASPER • deep red and opaque; strengthening to the body

CARNELIAN • soft orange color; soothing to the emotions

CITRINE • brilliant pale yellow; stimulating to the mind

AVENTURINE • speckled green; encouraging growth and expansion

BLUE LACE AGATE • soft lacy pale blue bands; gently soothing to the mind

BLUE TIGER'S EYE • deep blue with cat's eye reflections; magical and mysterious

AMETHYST • soft purple; soothing dreams

ROSE QUARTZ • soft pink; gently nurturing

LEPIDOLITE • sparkling pink; for inner security

CLEAR QUARTZ • transparent, smooth; clearing the air

HAVING FUN WITH CRYSTALS

Children can easily be inspired to play with stones, and there are many different ways to start. Take a large group of multicolored tumblestones and suggest laying them out in a rainbow pattern or a spiral shape. Match stones together in groups of similar colors, then look at them closely and see how they are different—some opaque, some transparent, some shiny, some with stripes or specks of different colors. Find some interesting colorful magazine pictures of different scenes in nature— mountains, the ocean, the beach, or close-ups of flowers—and suggest children match crystals to the colors in the images. Talk to children about the stones they like and ask them to say why they enjoy them. These ideas will ignite the instinctive response children display to crystals, and the stones will stimulate their senses of touch and sight. They may decide to keep a special stone with them, and it can be interesting to read up about it and look at its properties to see why they may have chosen it. Crystals encourage children's creativity and increase their sensitive connection to the earth. In later life, they may even become geologists or crystal healers.

Crystals as power objects

For thousands of years human beings have collected and kept particular objects safe because they were considered to have protective or magical powers. This practice spans many cultures, and archeological examples of ancient protection objects are found all over the world. Some of these may be of animal or plant origin, but many are stones or crystals. Amber, for example, has been found in prehistoric burial sites laid out on bodies, obviously intended as a symbol of significance.

Human beings always seem to have had a sense of powers and forces at work beyond rational understanding. Collecting particular objects in ancient times may well have been a way of trying to keep those powers at bay, surrounding the person with a layer of protection. Human belief is very powerful, so occasionally sacred objects were passed down the generations and credited with amazing abilities. It is easy to dismiss this practice as mere superstition, but it has persisted throughout our history.

LEFT *For the Ancient Egyptians, the scarab beetle was an important symbol; it represented Khepri, the god of rebirth and renewal. These amulets were carried by rich and poor as protection.*

GOLD

RED JASPER

JET

JADE

LAPIS LAZULI

PROTECTIVE ANCIENT AMULETS

The Ancient Egyptians, for example, became famous for the manufacture of amulets during the 2,500 years of their history prior to the Roman era. Amulets were small carved figurines depicting gods, symbols, or animals. Higher-ranking Egyptians could afford amulets made of gold, silver, or precious stones, but the use of amulets soon spread throughout the population with cheaper versions made of clay fired with layers of enamel. Amulets had a wide variety of uses: to bring good luck, to confer power, to add to funeral offerings, to offer protection and healing, and, when carved in the image of deities, to worship the gods. Some of the most famous Egyptian amulet symbols are the *ankh* (symbol of life), the *udjat* (the symbolic eye of Horus) and the scarab (a sacred beetle, symbol of Khepri, the god of rebirth). These were often carved in semiprecious or precious stones and placed on mummified bodies in tombs to protect the spirit of the deceased.

Another time in which amulets were very popular is early medieval Europe, between the eleventh and thirteenth centuries CE; the noble classes could afford them as engraved signet rings, cameos, or engraved gems set in gold or silver to act as protection. Cheaper stones, including jasper, were believed to safeguard women in childbirth. Precious stones such as spinel or sapphire were credited with protective, even life-saving powers, and battles were fought to possess them.

MODERN POWER OBJECTS

What do power objects mean to us in modern times? Looking back at the Ancient Egyptian uses of amulets for protection, healing, good luck (positive energy), and to invoke the beneficial energy of gods (the spiritual world), it is easy to see parallels with healing intentions today. The language may be different, but the focus is similar. Perhaps the main difference is the level of belief in the power of a crystal. Ancient people thought the stone itself had power; the modern way to interpret this is to say that crystals carry particular energy frequencies, and when someone is attracted to a stone, it means that the person has a strong affinity with or need for that kind of energy, which in turn will revitalize their system.

Wearing crystals for healing

Wearing jewelry is something many of us take completely for granted; something we do to accessorize our clothes, something that makes us feel good. However, do we ever stop to consider what we choose to wear as jewelry and why? Might there be a deeper significance behind the kinds of jewelry we choose? In the Bible, in the book of *Exodus*, God gives instructions for the manufacture of a breastplate for his high priest, Aaron. It contains precious stones such as topaz, agate, emerald, diamond, sapphire, amethyst, and beryl. This account shows that these precious stones together denote the rank and status of the high priest, and also point to his spiritual power. Religious leaders, kings, queens, princes, and potentates of the ancient world wore special jewelry not just for show, but to demonstrate their role as intermediaries with the spiritual realms. Such cultures as the Ancient Egyptians even decked out their high-ranking dead in these items of ritual jewelry so they would be recognized in the next life.

This may seem a long way from modern life, but we do have access to an extremely wide selection of precious and semiprecious stones these days, and they are available to everyone. Study and appreciation of their qualities can lead to a greater awareness of why they are being worn.

JEWELRY CRYSTALS

Here is a list of some typical crystals that may be chosen as jewelry for energetic reasons:

ROSE QUARTZ • to enhance unconditional love

CITRINE • to energize the mind

IOLITE • to inspire creativity

AMETHYST • to relax the mind

LAPIS LAZULI • to expand the third eye

AMBER • to enhance reproductive energy

MOLDAVITE • to encourage transformation

CLEAR QUARTZ • to keep a clear focus

MOONSTONE • to balance the hormones

CLEAR QUARTZ

Wearing crystals is a convenient and simple way to enjoy their effects. Set in gold or silver, precious and pure metal elements that contain the crystalline energies perfectly, these crystals are no longer simply ornamentation. They are healing tools that enhance and revitalize the energy field of the wearer. This is why sometimes it does not feel right to wear crystal jewelry that has belonged to someone else; that person's energy frequency will still be around it. If the piece is cleaned properly and reprogramed (*see pages 282–83*) then it will feel different and can be worn by another person.

ABOVE Wearing crystals brings their energy directly into contact with the skin; they bring pleasure because they are beautiful objects and have healing effects.

The most effective type of jewelry to choose for healing purposes is a pendant that hangs down toward the heart. The heart is a bridging chakra between spirit and matter, and combinations of crystals work well there to spread their energy throughout the auric field. Crystals can also be placed in a small pouch attached to a cord and worn around the neck, hanging over the chest to keep special stones close to the heart.

Crystal Healing

There are many methods of using crystals for healing purposes, and some of these have already been discussed, such as meditation with crystals, placing crystals in the bath, or making and using crystal remedies. This section covers the placement of crystals on or around the body, which is an ancient healing art. It involves working with another person, using crystals to balance, ground, or cleanse the energy frequencies in the body. Three types of layout are offered as a way to explore how crystals feel when they are placed in the energy field—the recipient simply relaxes, lying down to receive the treatment.

As with any healing approach, it is important to consider the recipient's needs carefully and to listen to any feedback given. Trust and communication between giver and receiver are vital. Another key aspect to working with crystals is intuition. If a strong impulse suggests placing a crystal in a particular location, it is worth listening to this because it indicates the receiving of information that a certain tool is required in that place. Many people are not used to listening to intuition, and the more work that is done with crystals, the more this ability is developed—it lies within everyone.

Chakra-balancing layouts

Throughout this book, crystals have been linked to the main chakra centers as well as color variations connected to higher levels of subtle energy. This first suggested layout uses crystals in the main chakra color bands to rebalance and nourish each one of the main centers. This brings a sense of deep relaxation and well-being that may be experienced as tingling energy, warmth, or mental impressions. All of these are useful feedback to the person placing the stones.

AMETHYST

LAPIS LAZULI

BLUE LACE AGATE

AVENTURINE

CITRINE

SUNSTONE

RED JASPER

PREPARING A CRYSTAL-HEALING SESSION

Setting up a clear and clean space is an important part of crystal work. Dust, untidy piles of magazines, and general clutter are not conducive to healing. Make sure the space is clean, bring in some fresh flowers in a vase, light some candles, and lay a thick mat on the floor so your friend can be comfortable lying on his or her back. Put a clean white sheet over the mat and have some cushions to hand—one under the head and one under the knees helps the back to relax. A soft shawl or light blanket can be helpful to keep the person warm. It is not necessary to undress for crystal work—as soon as the body relaxes its temperature can decrease.

Select the crystals that are needed for the layout (small tumblestones, points, or clusters are ideal). Place them in a glass bowl and run them under cold water for a minute to cleanse them. This is particularly important if they are used frequently for healing. It is a good idea to cleanse them again after a treatment. It is a matter of personal preference whether you choose to keep some crystals for your own use and a separate set for healing work with others. Do what feels right to you.

Then place the stones as directed on the next page to create the healing layout, and let your friend rest for at least fifteen minutes, breathing slowly and calmly.

CHAKRA CENTERS

Before showing the crystal selections, here are the seven main chakra centers with their locations (where stones are placed) and colors.

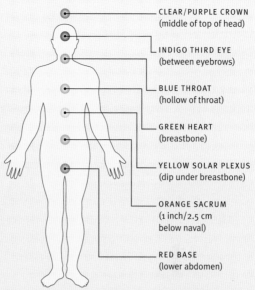

CLEAR/PURPLE CROWN
(middle of top of head)

INDIGO THIRD EYE
(between eyebrows)

BLUE THROAT
(hollow of throat)

GREEN HEART
(breastbone)

YELLOW SOLAR PLEXUS
(dip under breastbone)

ORANGE SACRUM
(1 inch/2.5 cm
below naval)

RED BASE
(lower abdomen)

CHAKRA LAYOUTS

LAYOUT 1

This is a gently balancing and harmonizing choice of stones which contain lovely complementing colors.

CROWN • amethyst (place on the mat above the crown of the head)	
THIRD EYE • lapis lazuli	
THROAT • blue lace agate	
HEART • aventurine	
SOLAR PLEXUS • citrine	
SACRUM • sunstone	
BASE • red jasper	

LAYOUT 2

This layout brings in some of the higher-energy-frequency crystals, and is suitable for someone who practices regular meditation because it enhances spiritual awareness.

CROWN • phenacite	
THIRD EYE • iolite	
THROAT • blue moonstone	
HEART • kunzite	
SOLAR PLEXUS • golden topaz	
SACRUM • amber	
BASE • ruby	

In both cases, it helps to place a piece of smoky quartz on the mat between the feet; if it is a pointed piece, have the point facing downward. This acts as an anchor and sends negative energy into the earth for neutralizing.

Grounding crystal layouts

Being grounded can be a challenge in the modern world, where our minds and senses are constantly bombarded with stimulation from all quarters. It is easy to live completely inside our heads and to forget the body and its needs for relaxation and restoration. These needs can then surface as signs of "dis-ease," or lack of ease, taking the form of headaches, muscular tension, colds and influenza, backache, digestive problems, or any number of other ailments. These are all physical signs of the effects of stress on the body.

GETTING IN TOUCH WITH THE EARTH

Being grounded allows you to return to a state of harmony with the earth. Until relatively recently in our history, our ancestors lived much more

AMETHYST

closely in tune with the seasons and the cycles of nature; before electric light, dawn and dusk dictated when people got up and went to sleep. Modern life has separated many human beings from the ground on which they stand. Using crystals helps to reestablish this connection, which is a deep and basic need.

Crystals come from deep within our planet; they have grown and formed over millennia through the powerful cycles of heat and cold, movement and stillness of the earth. Their energy frequencies create a sense of nurturing and deep comfort, a safe place to rest in the earth's energies. In particular, these grounding layouts bring a sense of deep peace. They restore body and mind and strengthen the energy field, countering the effects of stress.

As explained on pages 298–99, clean and set up the healing space and lay a mat on the floor covered with a white sheet. Position cushions so that your friend is lying comfortably on his or her back. Rest in the layouts for at least fifteen minutes.

MOSS AGATE

MALACHITE

ABOVE AND LEFT *Combining amethyst to soothe headaches, moss agate to strengthen the aura and malachite to gently warm the system will help ground and center the body.*

SMOKY QUARTZ GRID

A grid is a pattern of crystals laid on the ground around the body, creating a field of crystalline energy in which to rest. This is a simple pattern that involves placing eight small tumblestones of smoky quartz in an oval formation, one over the top of the head, three spaced equally down each side of the body (by the shoulders, waist, and knees) and one between the feet. Smoky quartz is one of the most effective grounding crystals, and it also protects the body from environmental stress.

RIGHT *A crystal grid can be useful if a person feels vulnerable or sensitive to their environment— it is calming and protective.*

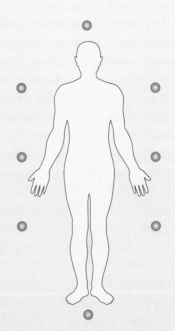

GROUNDING LAYOUT

GROUNDING AND NURTURING CRYSTAL LAYOUT

This involves placing crystals on the main chakra locations in a combination which is designed to achieve a grounding and peacefully nurturing effect. If you do not have all the crystals listed here, have a look at others in the same color group and select from that category.

CROWN • amethyst (place on the mat above the top of the head)

THIRD EYE • sapphire

THROAT • angelite

HEART • pink-and-green fluorite

SOLAR PLEXUS • amber

SACRUM • carnelian

BASE • hematite

Also add a piece of black tourmaline between the feet to act as an anchor to the earth and neutralize any negativity. After the treatment, exchange feedback with your friend about their experience in the layout.

Spiritually expansive crystal layouts

These combinations of crystals are designed to energize the chakras and expand the spiritual awareness of the receiver. The layouts can enhance psychic ability and open sensitivity to levels of consciousness beyond daily reality. These processes should only be explored once balancing and grounding layouts have been performed a few times; this is because the energy system needs to be functioning in a strong and harmonious way to act as a firm base for spiritual expansion to take place. These things cannot be forced; in ancient mystery, school pupils studied for years to attain the internal energetic balance needed for this kind of work. Sometimes in this day and age there is an impatience to have it all at once, but proceeding without caution can lead to energy imbalance in the body, even oversensitivity to one's surroundings.

DOWSING—A DIRECT LINK TO INTUITION

The best way to decide if one of these layouts is appropriate is to use a pendulum. Dowsing, as this is called, is a way of obtaining yes and no answers beyond the conscious mind, by using an instrument that interacts with your subtle energy field and, by extension, the universal field. It takes practice, but if you stop trying too hard with the rational mind it usually works easily. A pendulum can be made from any heavy object evenly balanced on a chain; however, these days crystal pendulums made of specially shaped quartz are popular. Clear quartz is good for obtaining clear answers. Hold the pendulum in your dominant hand and ask it first, "Please show me yes," and then, "Please show me no." Many people get a clockwise circle for yes and a straight right-to-left line for no, but there are variations. Test it by saying your real name and then a false name. Once you have set up a clear yes or no you can use the pendulum to test whether the expansive crystal layouts are for you or your friend at this time. If not, then test to see whether you need to continue with balancing or grounding work first. Always ask questions with yes or no answers.

Set up the healing space as directed on the previous pages. Let your friend rest for at least fifteen minutes in the layout and then share feedback on the experience.

LEFT *Use the pendulum to test which layout is right for you or your friend by asking, "Is the dynamic layout suitable for ... at this time?" Then repeat the question for the gentle layout.*

LABRADORITE

CITRINE

EXPANDING LAYOUTS

DYNAMIC LAYOUT (ACTIVE)

The dynamic layout is powerful and works quickly, using transformative crystals.

CROWN	• Herkimer diamond
THIRD EYE	• labradorite
THROAT	• sapphire
HEART	• moldavite
SOLAR PLEXUS	• citrine
SACRUM	• sunstone
BASE	• ruby

GENTLE LAYOUT (RECEPTIVE)

The gentle layout is more calming and slow in its action.

CROWN	• clear quartz (pointing upward)
THIRD EYE	• iolite
THROAT	• turquoise
HEART	• green jade
SOLAR PLEXUS	• amber
SACRUM	• carnelian
BASE	• red tiger's eye

Anchor both layouts by putting a piece of rutilated smoky quartz between the feet.

Professional crystal healing

Consulting a professional crystal healer is often a natural step for people who have become interested in collecting or carrying crystals themselves. Therapeutic treatment gives a deeper experience of what crystals have to offer. A professional crystal healer based in London comments, "People like crystals, often finding they are attracted without knowing why. Perhaps they like the colors ... maybe someone bought or gave them a piece, they've carried it and felt it do something for them." These are often the main reasons why people consider having crystal healing.

EXPERIENCING A SESSION

Crystal healing is a complementary therapy available on a privately funded basis. Consultations are one-to-one with the therapist, who usually sets up a treatment room with a couch for the client to lie on. Clients are encouraged to wear comfortable clothing, but they do not undress. Generally, crystals are placed on the body, although this is not absolutely necessary, only if the client is willing. A first consultation involves the therapist taking details of the client's medical history, then asking why they have come and what they hope to get from the session. The London-based crystal healer mentioned above likes to show crystals to her clients, even suggest that they touch them. She says, "Sometimes people want to know what crystals are or how they work, and others just want to relax and let me get on with it!"

She works very intuitively, laying crystals on the body or around it; the aim is to balance the energy system first, before any deeper work is undertaken. Clients give different kinds of feedback. Some feel sensations of warmth or cold, tingling, sweeping sensations over the body, or sometimes the location of an energy block that needs clearing. Some people see colors, hear sounds, or connect to other levels of awareness. Other people do not really feel anything, possibly because they have very analytical minds, but the energy of the stones is working anyway. Sometimes, after the treatment, this crystal healer gives a stone to a client to carry for a while, to keep the energy of the session working until that level has been cleared.

INTUITION, HEALING AND BALANCE

The crystal healer works intuitively, but says there are many approaches, some of which are much more structured and link stones to chakras in specific ways. She is a firm believer that clients will be drawn to the approach that is most suitable for

ABOVE *Experiencing the power of crystals in a healing session is deeply relaxing, bringing inner balance to body, mind and spirit.*

them. Crystal healing brings benefits, including balancing energy centers and pathways, deep relaxation, hormone and endocrine balance, and physical or emotional clearing within the body. Work on a specific issue might require five or six sessions to make a difference. The crystal healer comments: "Crystals are subtle yet powerful tools. Not everyone feels their energy in the same way. You need to let go of expectations and accept the stones, intuitively get to know them and trust what you absorb from them. Crystals are part of us; we come from the same source. One session can change your life!"

Crystals for physical health

Many crystals are considered to help particular physical conditions. The associations are sometimes linked to ancient healing traditions, or to more modern uses of crystals in healing practice. It is important to stress that keeping, holding, or using a crystal associated with a condition is no replacement for conventional medical advice. Any persistent conditions or changes in physical symptoms should be reported to your doctor. However, having an appropriate crystal with you allows its energy to support you in your healing process.

HEALING CRYSTALS

ALLERGIC REACTIONS • hematite, bloodstone, rose quartz

ANEMIA • garnet, bloodstone, hematite

ASTHMA • malachite, rhodochrosite, chrysocolla

BACKACHE • amber, citrine, danburite

BACTERIAL INFECTIONS • sulfur, malachite, aventurine

BLADDER PROBLEMS • prehnite, amber, orange calcite

BLOOD CIRCULATION • red jasper, bloodstone, hematite

BLOOD PURIFICATION • ruby, garnet, malachite, aventurine

BOWEL PROBLEMS • lepidolite, peridot, green fluorite

BROKEN BONES • calcite (any color), azurite, fluorite (any color)

CANCER SUPPORT • rose quartz, sugilite, watermelon tourmaline

CHRONIC FATIGUE SYNDROME • ruby, amber, aventurine, amethyst

CONSTIPATION • smoky/rutilated smoky quartz, black tourmaline

DETOXIFICATION • smoky quartz, sulfur, malachite

DIGESTIVE PROBLEMS • green fluorite, amber, citrine

EATING DISORDERS • rose quartz, kunzite, watermelon tourmaline

ENVIRONMENTAL POLLUTION • smoky quartz, zincite, merlinite

EYE PROBLEMS • aquamarine, blue chalcedony, blue tiger's eye

FERTILITY (to improve) • rose quartz, garnet, blue moonstone

FIBROMYALGIA • amethyst, aventurine, blue lace agate

FOOT PROBLEMS • black onyx, smoky quartz, prehnite

HEADACHES • rose quartz, blue lace agate, aquamarine

HEART ORGAN ISSUES • lavender quartz, rhodochrosite, garnet

The list below is by no means exhaustive; it suggests possible crystals in each context. The ways to use these stones are those already featured in this book, such as holding, carrying, or wearing a crystal, making a crystal remedy to sip during the day, or using a crystal in the bath. Any crystal can also be placed over the relevant part of the body and left there for about fifteen minutes to feel its effect. If meditating with a particular crystal, ask to be shown any relevant information to the condition you are experiencing. Repeat any treatment as needed.

HIGH BLOOD PRESSURE • amethyst, chrysocolla, blue chalcedony

HORMONE BALANCE • amber, blue moonstone, carnelian

IMMUNE SUPPORT • Herkimer diamond, aventurine, rutilated clear quartz

JOINT SORENESS • green calcite, azurite, rhodonite

KIDNEY SUPPORT • aquamarine, orange calcite, smoky quartz

LIVER SUPPORT • carnelian, red jasper, charoite

LUNG SUPPORT • lapis lazuli, turquoise, rhodonite

MENOPAUSE SUPPORT • blue moonstone, carnelian, amber

MENSTRUAL SUPPORT • bloodstone, white moonstone, carnelian

MUSCLE ACHES • hematite, danburite, red jasper

OSTEOPOROSIS • calcite (any color), fluorite (any color), azurite

PREMENSTRUAL SYNDROME • amber, blue moonstone, kunzite

REPRODUCTIVE ORGANS (f) • chrysoprase, blue moonstone, golden topaz

REPRODUCTIVE ORGANS (m) • jade, chrysoprase, carnelian

SCIATICA • jade, lapis lazuli, amethyst

SKELETAL SUPPORT • amazonite, fluorite, pyrite

SKIN ISSUES • rose quartz, azurite, citrine

SORE THROAT • angelite, celestite, blue lace agate

SURGICAL RECOVERY • amber, rose quartz, chrysoprase

TEETH PROBLEMS • calcite (any color), selenite, fluorite (any color)

THYROID SUPPORT • citrine, aquamarine, lapis lazuli

VARICOSE VEIN SUPPORT • amber, bloodstone, snowflake obsidian

WATER RETENTION • hematite, citrine, smoky quartz

Crystals for psychological health

This list of crystal correspondences links to moods, feelings, and mental attitude. The use of crystals to soothe the mind is another ancient practice; thousands of years ago, in India, particular crystals were linked to deities and accredited with different mental powers. Modern crystal healing suggests that psychological associations with crystals are individual; crystals are attractive because of their color, light reflections, or shape, and tend to represent levels of energy that are missing in a person's life, which they can then replenish.

CRYSTALS AND THE MIND

ADDICTION RECOVERY • malachite, peridot, smoky quartz

ANGER • rhodonite, chrysocolla, lapis lazuli

ANXIETY • rose quartz, amethyst, aventurine

ASSERTIVENESS (to encourage) • citrine, yellow tiger's eye, sodalite

ATTENTION DEFICIT DISORDER • kunzite, selenite, azurite

BREATHLESSNESS (stress) • amber, seraphinite, green calcite

BURN OUT • ruby, garnet, carnelian

COMMUNICATION ISSUES • turquoise, lapis lazuli, angelite

CONFIDENCE (lack of) • citrine, amber, orange calcite

CONFUSION • fluorite (any color), charoite, lepidolite

COURAGE (to build) • bloodstone, hematite, yellow tiger's eye

CREATIVITY (to encourage) • iolite, phenakite, cathedral quartz

DEPRESSION (to ease) • amethyst, angelite, kunzite, pink tourmaline

DREAMS (to calm) • rose quartz, blue lace agate, prehnite

DREAMS (to encourage recall) • Herkimer diamond, labradorite, elestial quartz

DRUG ABUSE RECOVERY • smoky quartz, jet, rutilated clear quartz

EATING DISORDERS • rose quartz, kunzite, watermelon tourmaline

EMOTIONAL DISTRESS • lavender quartz, carnelian, purple fluorite

FATIGUE • amethyst, aventurine, seraphinite

FEAR • angelite, kyanite, lepidolite

FOCUS (to build) • fluorite (any color), calcite (any color), selenite

FORGIVENESS • kunzite, rose quartz, chrysoprase

If you are experiencing a feeling or a mood and a particular crystal jumps out at you, then understand that this is the one that is suited to your current needs. Follow your impulse and allow that crystal's energy to interact with yours. It should also be said that crystals are not necessarily a miracle cure, and if you are experiencing repeated emotional patterns and feelings of unhappiness then it is important to seek professional guidance. Crystal energies will support you in your healing journey; carry them, wear them, or make a crystal remedy to take.

GRIEF • rainbow obsidian, sugilite, rutilated smoky quartz

GROUNDING • jet, smoky quartz, black tourmaline

GUILT (to ease) • amazonite, aventurine, rose quartz

HYSTERIA • obsidian, hematite, smoky quartz

IMPOTENCE/LACK OF SEXUAL INTEREST • zincite, ruby, carnelian

INFERTILITY (emotional support) • pink tourmaline, ametrine, purple fluorite

INSOMNIA • amethyst, lavender quartz, iolite

IRRITABILITY (to calm) • angelite, blue chalcedony, lepidolite

JEALOUSY (to ease) • peridot, prehnite, citrine

JOY (to increase) • kunzite, sunstone, goldstone

LONELINESS (to ease) • carnelian, amber, orange calcite

LOVE (to bring into life) • ruby, rose quartz, emerald

MEDITATION (to improve focus) • lapis lazuli, iolite, Azeztulite

MEMORY PROBLEMS • green fluorite, amethyst, selenite

NERVOUS TENSION (to ease) • sodalite, watermelon tourmaline, carnelian

PEACE (to encourage) • angelite, seraphinite, pink sapphire

PROTECTION • bloodstone, spinel, hematite

PSYCHIC DEVELOPMENT • lapis lazuli, iolite, channeling quartz

RELATIONSHIPS (to improve) • kyanite, kunzite, peridot

RELAXATION (to increase) • rose quartz, green calcite, blue lace agate

SELF-ESTEEM (to improve) • citrine, amber, ruby

SHYNESS (to decrease) • sunstone, orange calcite, zincite

TRAUMA RECOVERY • aventurine, citrine, chrysocolla

Crystals as birthstones

Many cultures have linked crystals to particular months or aspects of astrology. These are called birthstone or zodiac lists, and there are many variations according to particular traditions. Birthstones are considered to bring good fortune if they are worn by someone with a birthday in a particular month or sign of the zodiac. However, these days birthstone lists are more often used as a marketing tool by jewelers to sell particular gems. There is no need to be tied to these lists when it comes to choosing a precious gem, crystal, or stone to wear; if you are drawn to a crystal instinctively, then it has relevance to you, and it does not matter where it falls on a birthstone or zodiac stone chart. However, if you want to select a gem or crystal to give as a gift, or if you want to wear something associated with your birthday or astrological sign, then these lists will be helpful to you.

CLASSIC BIRTHSTONE LIST

MONTH	BIRTHSTONE
JANUARY	garnet, rose quartz
FEBRUARY	amethyst, onyx
MARCH	aquamarine, bloodstone
APRIL	diamond, clear quartz
MAY	emerald, chrysoprase
JUNE	pearl, moonstone
JULY	ruby, carnelian
AUGUST	peridot, sardonyx
SEPTEMBER	sapphire, lapis lazuli
OCTOBER	opal, tourmaline
NOVEMBER	topaz, citrine
DECEMBER	tanzanite, turquoise

GARNET

DIAMOND

TURQUOISE

EMERALD

TOPAZ

PERIDOT

Wearing a birthstone is considered to bring good fortune to those who wear it.

ZODIAC SIGN LIST

Listed below are the sun signs of the zodiac, with a selection of crystals linked to these signs. If you possess a detailed horoscope chart done for your date, time, and place of birth, this gives accurate information about your astrological profile: it will highlight the position of planets and the moon, showing where they occur in different signs of the zodiac on your chart—moon in Scorpio, for example. You can use this information to wear crystals linked to the planet positions in your horoscope if you choose to do so.

Where a stone is listed that is found in different colors—tiger's eye, calcite, fluorite, or jasper, for example—simply use your intuition to choose which one works best for you. It is interesting to note that some crystals appear in several different signs because they have a wide range of energetic effects.

ZODIAC CRYSTALS

ARIES • (Mar 21–Apr 19)
carnelian, jasper, ruby, diamond, kunzite, bloodstone

TAURUS • (Apr 20–May 20)
aquamarine, tourmaline, topaz, emerald, tiger's eye

GEMINI • (May 21–Jun 20)
citrine, tiger's eye, chrysocolla, pearl, apophyllite

CANCER • (Jun 21–Jul 22)
pearl, moonstone, emerald, ruby, amber

LEO • (Jul 23–Aug 22)
sunstone, clear quartz, ruby, turquoise, spinel

VIRGO • (Aug 23–Sep 22)
carnelian, citrine, sapphire, peridot, sugilite

LIBRA • (Sept 23–Oct 22)
opal, lapis lazuli, peridot, aventurine, jade

SCORPIO • (Oct 23–Nov 21)
kunzite, Herkimer diamond, aquamarine, malachite, dioptase

SAGITTARIUS • (Nov 22–Dec 21)
smoky quartz, turquoise, malachite, spinel, blue lace agate

CAPRICORN • (Dec 22–Jan 19)
onyx, jet, ruby, garnet, labradorite

AQUARIUS • (Jan 20–Feb 18)
amethyst, aquamarine, angelite, blue chalcedony, sapphire

PISCES • (Feb 19–Mar 20)
calcite, turquoise, pearl, fluorite, rose quartz

AGATE: a type of quartz with a microcrystalline structure and distinctive bands

AGGREGATE: a rock composed of different minerals

ALCHEMY: ancient mystical science seeking to turn lead to gold

AMORPHOUS: a mineral with no repeating internal structure and an irregular shape (e.g. obsidian)

ASTERISM: a starlike light effect on a polished stone (e.g. ruby or sapphire)

AURA: term for the energetic field around the human body

BORYTROIDAL: shaped like a bunch of grapes

CABOCHON: shaped crystal with flat base and polished dome surface

CHAKRA: term describing one of the seven energy centers in the body

CHATOYANCY: light effect resembling a cat's eye

CLEAVAGE: angle in a crystal along which it will split

COLUMNAR: shaped like a column

CRYSTAL: mineral with geometrical internal structure creating angled planes and faces

CUBIC: in the form of a cube, a three-dimensional, four-sided structure

DICHROIC: a mineral showing two different colors at different angles

ELEMENT: a single chemical substance

ENCRUSTATION: a crystal that forms a crust over a base layer

FACET: angled crystal face, either natural (e.g. quartz) or man-made (e.g. cut ruby)

FELDSPAR: a large group of crystals formed of aluminum silicate

GNEISS: metamorphic rock with quartz and feldspar present

HEXAGONAL: six-sided

IGNEOUS: a rock produced by fire or volcanic activity

INCLUSION: a mineral contained within another crystal (e.g. rutile)

INORGANIC: in mineral sense, all minerals are inert chemical compounds

IRIDESCENCE: a play of rainbow colors

LABRADORESCENCE: a shimmering effect of contrasting color (e.g. as seen in labradorite)

MASSIVE: a large mineral formation, usually microcrystalline layers

MINERAL: a solid chemical compound

OCTAGONAL: eight-sided

OPALESCENCE: a rippling sheen of rainbow hues (e.g. as seen in opal)

ORGANIC: in mineral sense, a compound originally from a living source (e.g. amber)

PIEZOELECTRIC: carrying an electrical charge after squeezing or striking

PLEOCHROISM: change of density of color depending on angle of light

REFRACTION: light diverted at a different angle

ROCK: combination of minerals and crystals (e.g. granite)

SCHILLER: sparkle or shimmer effect at an angle (e.g. as seen in tiger's eye)

STRIATIONS: natural grooves along the surface of a crystal

TABULAR: crystals shaped like slabs

TECTONIC PLATES: sections of the earth's crust floating on the mantle

TERMINATION: pointed crystal tip

ULTRAVIOLET: ray of light at top end of spectrum, invisible to the human eye

RED TIGER'S EYE

The following is a list of organizations that can be contacted if you are seeking a qualified therapist in your area or if you are interested in training as a crystal healer.

USA CRYSTAL HEALING ASSOCIATIONS

Crystal Academy of Advanced Healing Arts
www.webcrystalacademy.com

Association of Melody Crystal Healing Instructors
www.taomchi.com

The Crystal Conference
www.thecrystalconference.com
Tel (+1) 802 476 4775

Amethyst Rose Healing Arts
www.amethysthealing.com
Tel (+1) 703 868 5834

CLEAR QUARTZ

UK CRYSTAL HEALING ASSOCIATIONS

ACHO
(Affiliation of Crystal Healing Organizations)
www.crystal-healing.org
Tel (+44) (0)7837 696 301

IACHT
(International Association of Crystal Healing Therapists)
www.iacht.co.uk
Tel (+44) (0)1200 426 061

Crystal Healing Federation
www.crystalandhealing.com
Tel (+44) (0)870 760 7195

ROSE QUARTZ

CRYSTAL SUPPLIERS (USA)

Peaceful Mind
www.peacefulmind.com

Mountain Gems and Healing Crystals
www.healingcrystals.net

Best Crystals
www.bestcrystals.com

SMOKY QUARTZ

CRYSTAL SUPPLIERS (UK)

Richard Scull
(via his website he sells worldwide)
www.crystal-planet.com

Sara Giller
www.crystalvine.co.uk

Mike Jackson
www.thecrystalman.co.uk

GREEN CALCITE

AUSTRALIAN CRYSTAL HEALING

Natural Therapy Pages Resource Directory
(allows a search for qualified crystal therapists by region)
www.naturaltherapypages.com.au/therapy/Crystal_Therapy

The Karyna Centre
www.crystalsoundandlight.com

Index

HERKIMER DIAMOND

317

PERIDOT

319

ACKNOWLEDGMENTS

JENNIE HARDING BA TIDHA MIPTI HNC has twenty years' experience as a healer working with various modalities including crystals, essential oils, herbs, crystal energy remedies, incense, and natural approaches to beauty. She is the author of fifteen books on this range of topics. Between 1992 and 2005 she was senior Essential Oil Therapeutics tutor at the Tisserand Aromatherapy Institute. She practises Jin Shin Jyutsu, a Japanese energy-balancing art and teaches it as self-help. She dedicates her energy to creating and sharing tools for personal transformation and self-awareness, with a firm belief in self-empowerment and preparation for planetary change.

I would like to dedicate this book to my niece Elizabeth and my nephew Joshua; they are the new generation of Earth custodians.

I would like to thank Richard Scull for his immense expertise, patience, and expert contribution to the preparations for this book—including the loan of many crystals for the photography—and Sara Giller, Mike Jackson, Paul Bastiani, Charlies Rock Shop, Crystal Reflection and Crystal Fantasy for their generous loan of crystals for the same purpose. Thanks also to Hazel Oatey for her fascinating insight into the work of a professional crystal healer, and to Hazel Songhurst at Ivy Press for her wonderful editorial support during this project.